Hamlyn all-colour paperbacks

Michael H

Fossil

illustrated by The Garden Studio
Pat Oxenham

Hamlyn · London
Sun Books · Melbourne

FOREWORD

The origin of man has fascinated thinking people for many centuries, indeed the folklore of most primitive peoples still contains suggestions as to how their tribe or ethnic group came into existence. Few people nowadays can accept the bible story of Adam and Eve as a literal record of the events that took place, the vast majority accepting Darwin's view that animals and men changed, or evolved, into different species through geological time. This book is an attempt to show how this evolutionary process has taken place with respect to man and his forebears.

The most concrete evidence for the evolution of man consists of the fossil bones of man-like creatures who preceded us during the past. From interpretations of these fossil bones, ideas can be formulated as to how our ancestors looked, stood, walked, used their hands and, in broad terms, behaved.

The number of specimens of fossil men that are in our possession is at once both large and small. As representatives of the populations of ancient man that must have existed, they form a small sample; but some hundreds of skulls and skeletons, and several thousands of teeth are known. This is the raw material of palaeoanthropology.

The story of fossil man is incomplete, but every year brings new specimens, new methods of investigation and new knowledge. Gradually the gaps are being filled so that the continuity of human evolutionary change is more clearly seen.

Man's insatiable curiosity about man will always be the spur that drives him to seek himself in the rocks of the past.

M.H.D.

Published by The Hamlyn Publishing Group Limited
London · New York · Sydney · Toronto
Hamlyn House, Feltham, Middlesex, England
In association with Sun Books Pty, Melbourne

Reprinted 1972

ISBN 0 600 00069 9
Phototypeset by Oliver Burridge Filmsetting Limited, Crawley, Sussex
Colour separations by Schwitter Limited, Zurich
Printed in Holland by Smeets, Weert

CONTENTS

THE RISE OF THE PRIMATES

Classification of man

Whatever views are held upon the origin of mankind nobody seriously doubts that man is an animal, and therefore shares the ability to move, eat, sleep, grow and reproduce. The simplest of comparisons will show striking resemblances in the skull, the limbs, the senses, the brain and the general arrangement of the internal organs. Indeed it is possible to assign man to one of the major sub-divisions of the animal kingdom, a class of vertebrates known as the mammals.

The mammals are warm-blooded animals that are characterized by their method of giving birth, their ability to suckle young, the possession of hair and of two sets of teeth, milk and permanent. They are also characterized by certain features of the circulation. Man can be assigned even further to an order of mammals called the Primates, a group which includes tree-shrews, tarsiers, lemurs, monkeys and apes.

Classification is based upon structure and can be continued until individual species are recognized. A species can be defined as a group of animals actually, or potentially, capable of interbreeding and producing fertile offspring which are like their parents.

The classification of modern man

Kingdom	Animalia (Animals)
Phylum	Vertebrata (Vertebrates)
Class	Mammalia (Mammals)
Order	Primates (Primates)
Superfamily	Hominoidea (Hominoids)
Family	Hominidae (Hominids)
Subfamily	Homininae (Hominines)
Genus	*Homo* (Man)
Species	*sapiens* (Modern Man)

For centuries species were thought to be immutable and 'specially created' according to the letter of the bible story. It was not until 1859 that Charles Darwin produced massive evidence to prove that species are capable of gradual change, or evolution, through time so that new species can arise. Darwin also showed the way in which the changes are brought about, the process that is called *natural selection*. In the competition of life those forms that are best adapted to their surroundings are at a considerable advantage and will tend to succeed and multiply. Similarly those forms that remain adaptable, changing when the surroundings change, will also tend to survive while other more specialized creatures perish.

The primates arose about seventy million years ago from the mammalian stem during the Cretaceous and the Palaeocene periods. Their survival to the present day is due to their remarkable adaptability.

The Origin of the Primate Stock from the Mammalian Adaptive Radiation

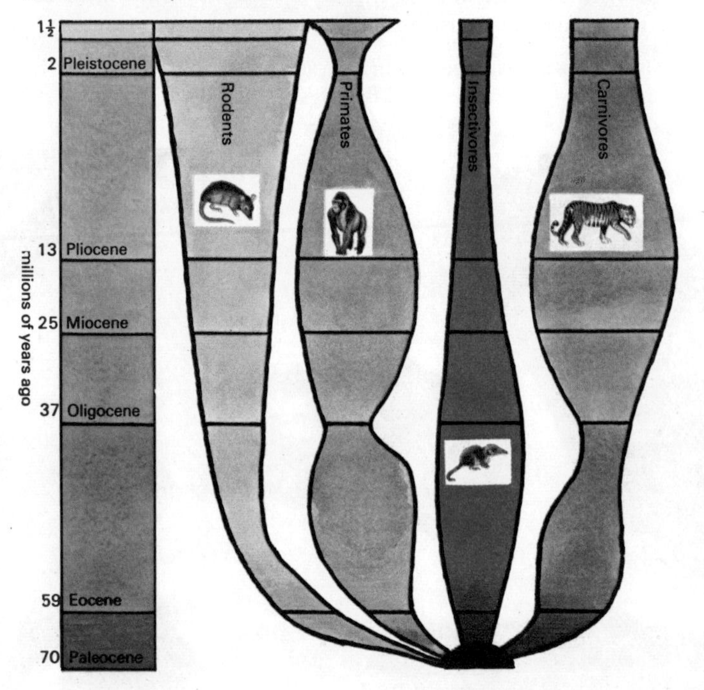

The primate radiation

What is a primate? What are the common features of this group of animals which lead us to link them together?

Nearly all primates live in trees or have had a near ancestor who lived in trees and this means that to be successful they must have a structure that will allow them to move freely in a three dimensional habitat. In general they must be agile and able to grasp branches as well as having good stereoscopic vision, which will therefore give them a good judgement of distance. Primates are generalized creatures that have kept an unspecialized form of the nands and feet. In addition thay have a collar bone, or clavicle, which directs the shoulder to the side of the body and widens the range of the upper limb. Flat nails help to support the finger pads and improve the grip.

The only weapons of the primates are their long dagger-like canine teeth. Their agility enables them to survive, however, because they are able to exploit the 'flight' rather than the 'fight' reaction towards an attack.

One feature that characterizes the primates is their ability to grasp objects. Flat nails and finger pads help to improve the grip.

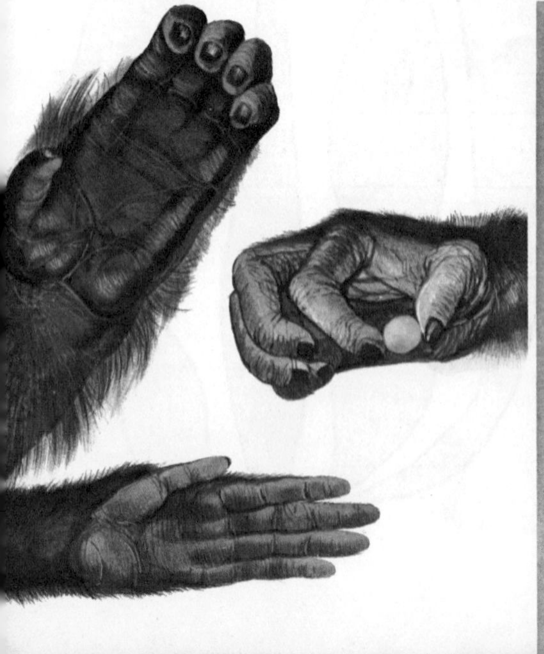

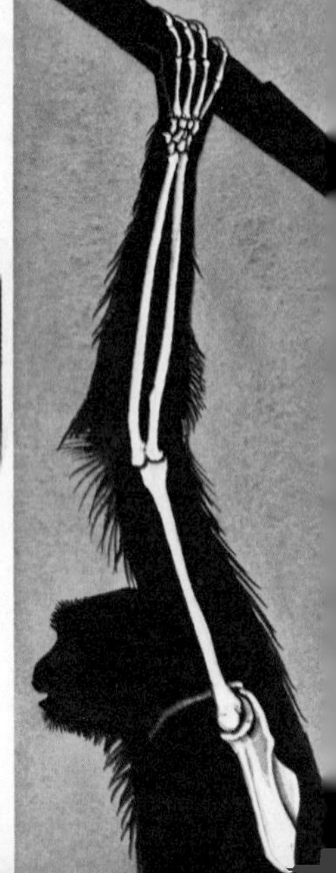

Once the forest was mastered, it was exploited by a wide range of primate varieties that spread and diversified until all levels were occupied, from the canopy to the shrub layer. Several times in evolution, groups of primates have come down from the trees to become ground-living animals during the times that the forests receded and became replaced by open savannah and grassland. It was one of these groups that formed the adaptive radiation which contained the earliest forerunners of man.

Tropical rain forest tends to be stratified and the different vegetation layers are occupied by primates that have different methods of locomotion.

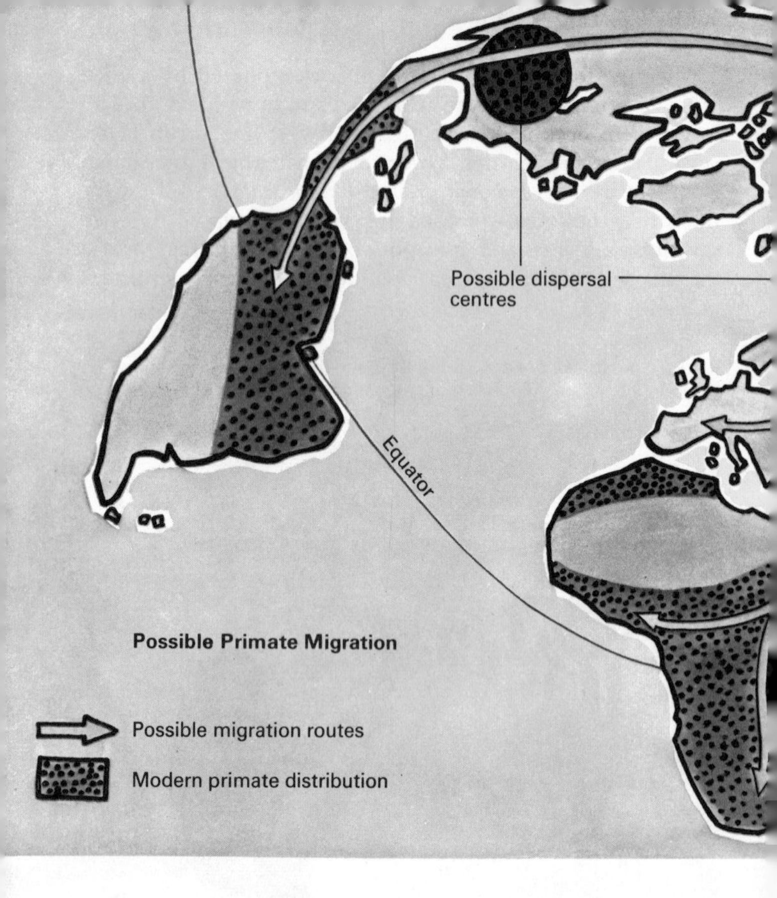

The spread of primates

The earliest primates were almost world-wide in distribution, and were small primitive animals of lemuroid or tarsioid form that spread and flourished in North America and Europe. These lived in trees and ate fruit and insects with little interference until the Eocene, when suddenly their numbers diminished. The advent of modern fast-breeding rodents and carnivores provided such competition for food and living space that many of the early generalized primates became extinct. In only a few places where they became geographically isolated, did these early forms survive. The lemurs of Mada-

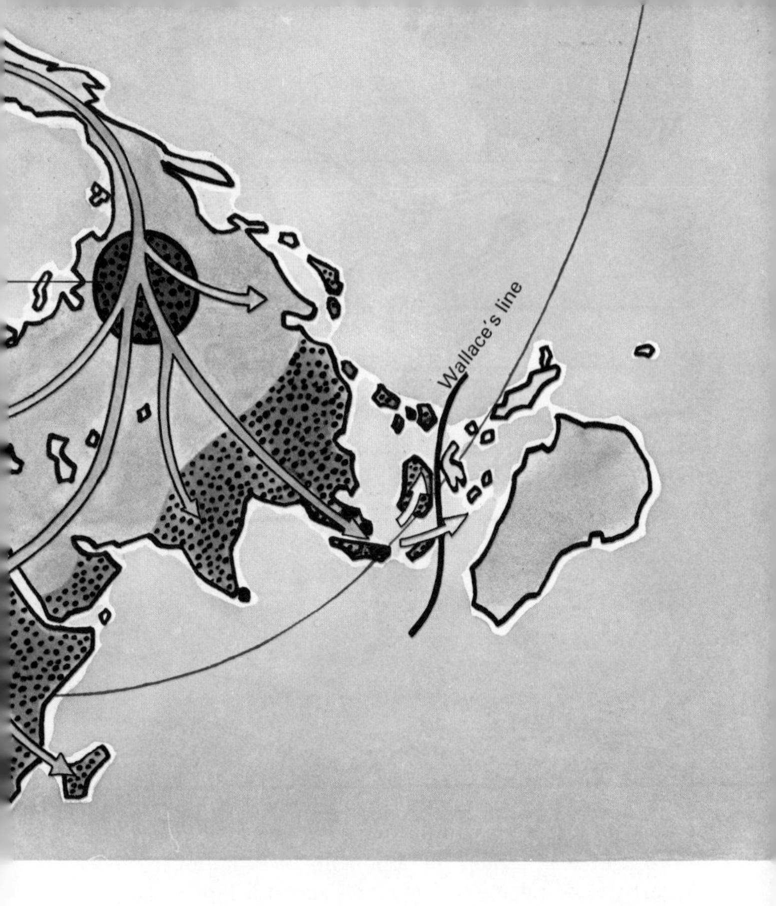

gascar, for example, are modern descendants of a stock that survived only because of their isolation on an island.

At this time north and south America were separate continents so that the early primitive primates, or prosimians, which reached South America probably did so by 'island-hopping' on drifting vegetation. From this stock came the radiation of New World monkeys that spread and proliferated in the Amazon forests. Little is known of the early history of the Old World monkeys except that primates did not penetrate throughout the Far East. There are no living or fossil primates, other than man to the East of Wallace's line.

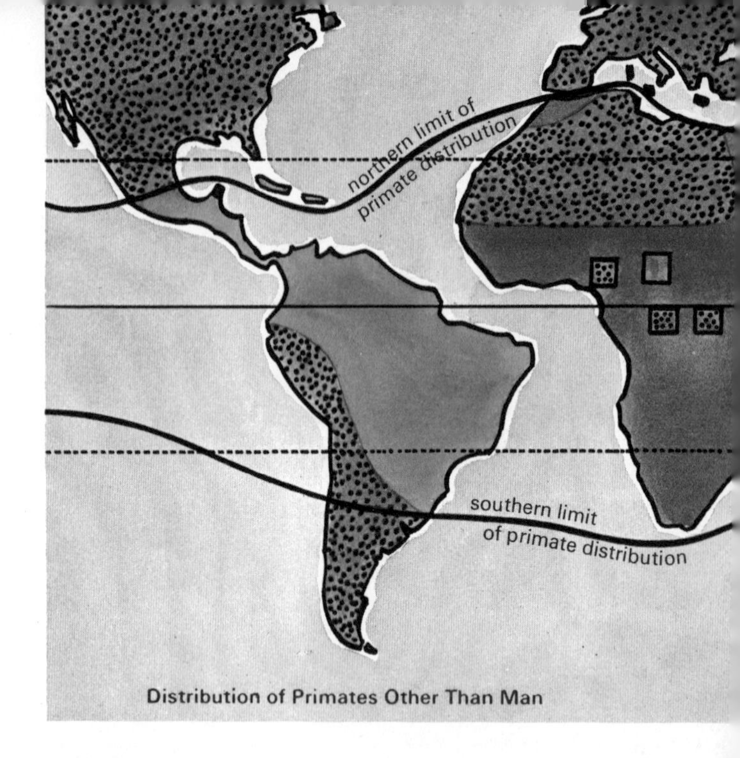

Distribution of Primates Other Than Man

Missing links

Perhaps the commonest fallacy that exists in the study of evolution is that 'man arose from the apes'. This misbelief has been used many times by opponents of evolutionary theory. Clearly no living ape can be man's ancestor but, if man's own evolutionary history is being considered, the history of other living primates must also be examined. There cannot be any 'missing link' in *modern* ape-like form.

From a common early primate stem, evolutionary advance has progressed towards greater complexity, culminating in the higher primates and man. Stages of advance are marked by points at which recognized groups diverged to pursue their own evolutionary fate. Some failed and became extinct. The relationship of the groups that are still living springs from the possession of a common ancestry. Man did not arise from the apes, but men and apes did arise from common primate stock.

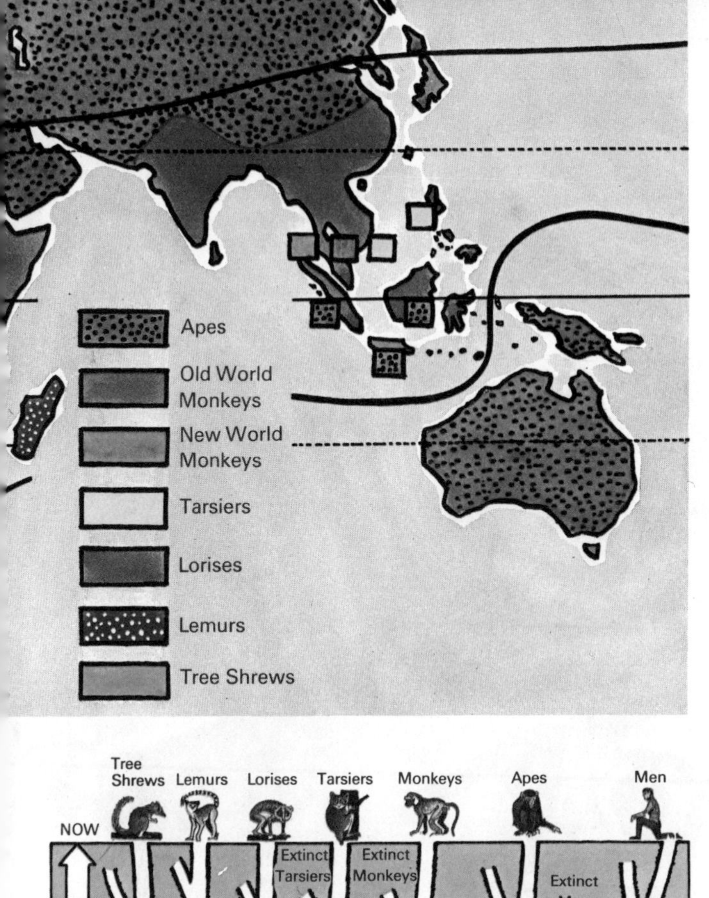

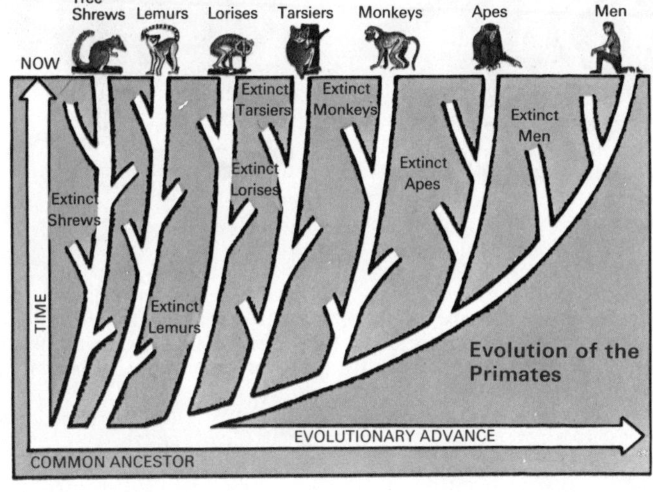

Evolution of the Primates

Modern primates

TREE-SHREWS (*Tupaia*) Small, squirrel-like primates from South-East Asia, west of Wallace's Line. These primitive primates have eye sockets surrounded by a bony ring and a second tongue (the sublingua) rather like lemurs.

TARSIERS A nocturnal group from South East Asia and the Philippines that have long legs and large hands with terminal finger pads. The large eyes enable them to see in dim light to catch insects and small lizards.

LORISOIDS A nocturnal group which includes the Slow Loris (*Nycticebus*), the Potto (*Perodicticus*), and the Bushbabies (*Galago*). The Galagos are found in Africa and the other lorises in Africa and the Far East. Galagos are active leapers while the other lorises are slow movers.

LEMURS Entirely confined to the island of Madagascar, the lemurs are gentle animals with almost no enemies except man. There are many varieties from the tiny mouse lemur (*Microcebus*) to the 'monkey-sized' *Indris*. The best known is the Ring-tailed lemur (*Lemur catta*).

NEW WORLD MONKEYS From Central and South America range from the little Marmoset (*Callithrix*) to the Spider Monkey (*Ateles*). Arboreal creatures, often with prehensile tails, living in tropical rain forests. The Squirrel Monkey (*Saimiri*), the Capucin (*Cebus*), and many others live on fruit and leaves.

OLD WORLD MONKEYS From Africa, India, and Asia, some live in trees and some on the ground. The Guenons, Mangabeys, the Baboons, the Macacques, the Colobuses and the Langurs, occupy a wide range of habitats and eat a variety of diets. Their tails are never fully prehensile.

APES From Central Africa and the Far East, this group is represented by the Chimpanzee (*Pan*), the Gorilla (*Gorilla*), Orang Utan (*Pongo*) and the Gibbons. All are tailess. Gorillas and Chimpanzees are often ground-dwellers while the Gibbons and Orangs are entirely arboreal.

MEN The most successful modern primate with a world-wide distribution, from the poles to the equator. Two-legged and large-brained. Has the power of speech, the use of tools and of fire, and the ability to control his immediate environment by means of clothes and shelter.

The early primate *Notharctus* may have looked like this

Early fossil primates

The earliest known fossil primates appear in deposits that belong to the middle of the Palaeocene Period of both North America and western Europe. Several primitive forms existed at this time but one of the most important groups was the Plesiadapidae. These were small squirrel-like quadrupedal animals that had forward-facing eyes, long faces and clawed fingers and toes. They appear to have been ground-living creatures but one cannot be certain that they did not spend some time in the trees. The dentition of these creatures is not unlike that of some present-day rodents, yet it seems most likely that these early primates were insectivorous.

During the Eocene Period the early primate stock proliferated into five or six groups. Two of these, the Adapidae and the Omomyidae, are of particular importance, because they possibly gave rise to both the lemurs and the old world monkeys and apes. Nearly all the fossil limb bones of Eocene forms suggest that they were active tree-dwellers whose grasping feet allowed them to cling to upright stems and trunks, while their long legs would allow leaping. The forward-facing eyes and overlapping visual fields of these early fossil primates would have been a great help in judging distances before jumping.

The Omomyidae seem most likely to be the stock from which the Oligocene primates arose in Africa towards the end of the Eocene Period. These forms are represented by a large group of fossils that were recovered from the Fayum district of Egypt. Most of these specimens consist of fragments of jaws and skulls only, but their examination has allowed the reconstruction of a provisional scheme of their relationships and possible descendants.

Three views of an *Adapis* skull

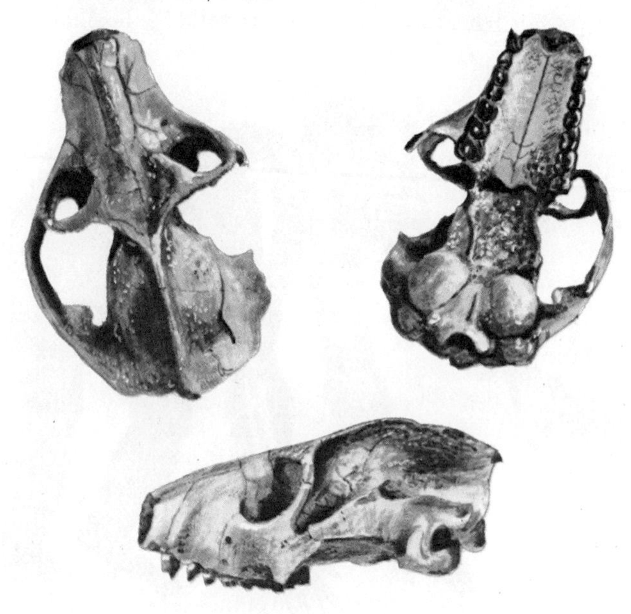

Fossil apes

The Miocene Period is remarkable for the number of specimens of fossil apes that have been recovered, while the deposits have yielded comparatively little evidence of fossil monkeys. The early apes belong to a well-known genus *Dryopithecus*, widely represented from Europe, Africa and Asia. Typical of these forms is a Y-shaped molar tooth fissure pattern.

A similar, if not identical, group from Africa are the Proconsuls, *Proconsul africanus*, *Proconsul nyanzae*, and *Proconsul major*. The teeth of *Proconsul africanus* are dry-opithecine but have some primitive features, while the limb bones suggest a type of locomotion not unlike that of the langurs of the far East.

Recent discoveries in Kenya have revealed new forms, *Kenyapithecus africanus* and *Kenyapithecus wickerii*. The earlier Miocene form, *Kenyapithecus africanus*, possibly has some advanced dental features but more evidence is needed to establish its relationship to the hominid line. The latter form, *Kenyapithecus wickerii*, bears some resemblance to *Ramapithecus*, a Pliocene ape whose dental arcade is remarkably hominid.

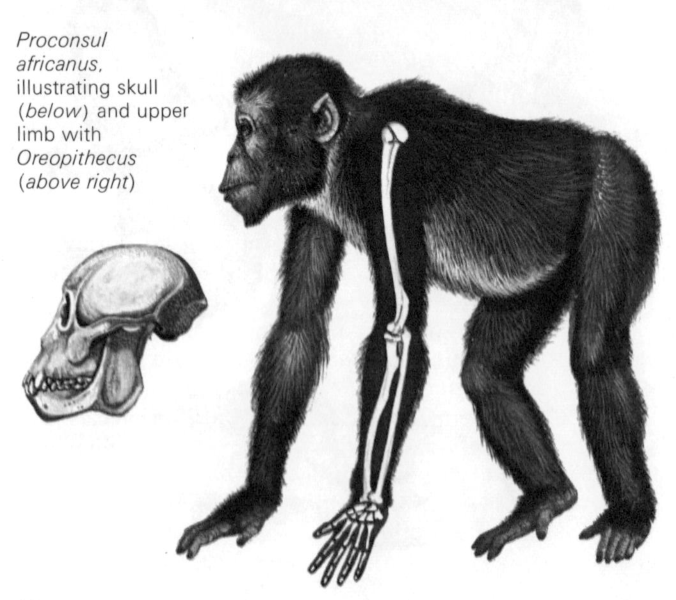

Proconsul africanus, illustrating skull (*below*) and upper limb with *Oreopithecus* (*above right*)

These two higher primates may well prove to be the earliest true hominids known, when more material becomes available for study. In particular, specimens of limb bones are needed, to establish their posture and method of locomotion. The only other significant remains from this Period belong to *Oreopithecus*, a small, highly arboreal form with long arms and specialized teeth that was recovered from the soft coal deposits of northern Italy.

At this stage there is a gap of about twelve million years in the knowledge of primate evolution since there are no more fossils available until the beginning of the Pleistocene Period. This Period was heralded by a wave of 'new' mammals, the Villafranchian fauna, many similar to those of today, but also including some giant forms. In general, Pleistocene primates are very similar to modern forms, but the most important ones are undoubted hominids and thus clearly related to mankind. The Pleistocene Period is the Period of fossil man.

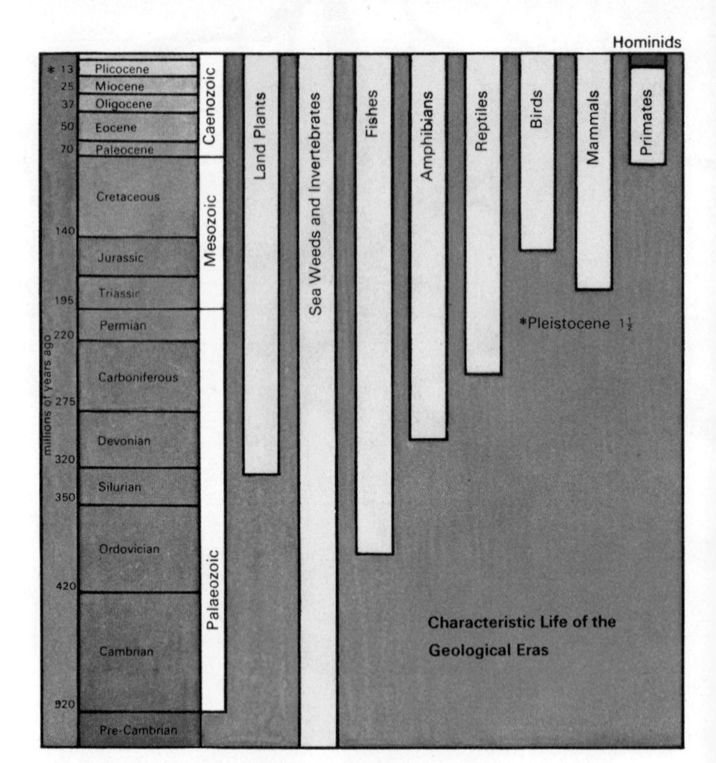

Man occupies a very small place in geological time.

THE PLEISTOCENE PERIOD

Pleistocene climate

All the remains of early man are attributed to deposits laid down during this geological epoch. It extended from approximately two million years ago to ten thousand years ago, the beginning of the Recent Period.

The evolution of man from his primate ancestry has taken place rapidly during the last two million years. One of the reasons for this is that during the Pleistocene Period the climate was very unstable. Arctic conditions spread down from the poles at least four times bringing snow and ice deep into Europe and North America. In Africa there were corre-

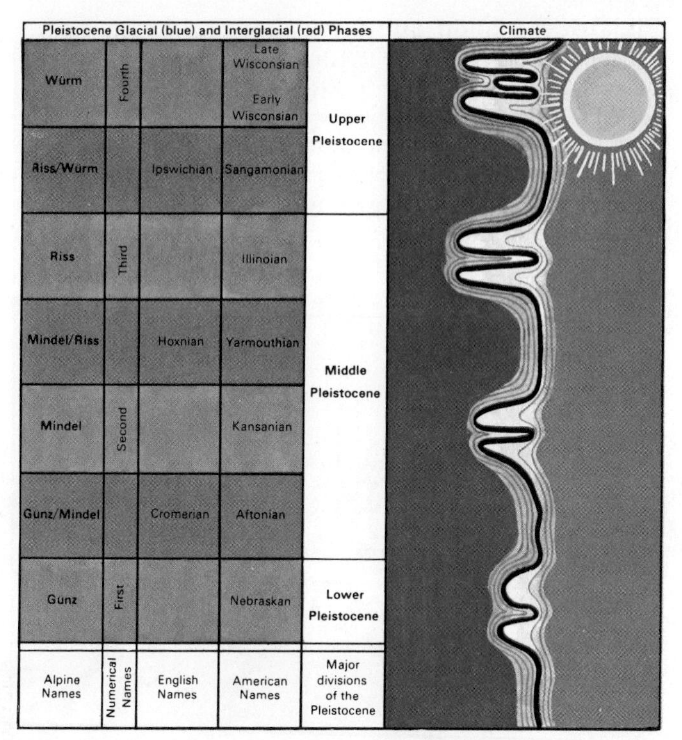

Pleistocene Glacial (blue) and Interglacial (red) Phases

Climate

Alpine Names	Numerical Names	English Names	American Names	Major divisions of the Pleistocene
Würm	Fourth		Late Wisconsian Early Wisconsian	Upper Pleistocene
Riss/Würm		Ipswichian	Sangamonian	
Riss	Third		Illinoian	Middle Pleistocene
Mindel/Riss		Hoxnian	Yarmouthian	
Mindel	Second		Kansanian	
Gunz/Mindel		Cromerian	Aftonian	
Gunz	First		Nebraskan	Lower Pleistocene

Approximate correlations of glacial and interglacial terminology.

sponding wet phases, of high rainfall. Thus, the evolutionary pressure on the early hominids was high and the environment ruthlessly selected out those forms which could not adapt to the changing conditions. There was, therefore, a high premium on any new and advantageous structural and behavioural characters, such as the ability to walk upright and use the hands for tools and weapons.

Glaciations dominated the climate during the last two million years and deposits left by the retreating ice indicate that four major cold phases occurred, each glaciation separated by an interglacial period. Smaller retreats are known as interstadials. Glaciations and interglacials are named variously in different parts of the world and can be provisionally correlated.

Pleistocene geography

During the glaciations, the ice sheets and glaciers of the polar and mountainous regions locked up vast quantities of the world's water so that sea levels were low. Great Britain was at that time joined to the European land mass and other land bridges formed in South East Asia. Land masses were depressed by the weight of the ice and recovery only occurred slowly as the ice melted in interglacial phases. This process is still

continuing in Scandinavia where the retreat of the ice cap over Finland is causing a rise of the land mass of approximately eleven millimetres each year. Later melt-water flowed into the sea and raised the sea level so that raised beaches, submerged forests and drowned valleys were produced. Rivers were slowed by the rising sea and broad swamps and deltas were formed as the silt carried by the stream was deposited. Later, when the sea level fell, downcutting of the river beds left river terraces on either side.

Streams from the melting ice front brought down mud and gravels into the periglacial tundra. As this region dried, winds raised dust storms and carried loess into the surrounding grassland steppe regions.

At the edge of a glacier, snow melt produces water at the margin of the ice and beyond it (*left*), as well as depositing material carried by the ice. Fluctuations of sea and river levels during glaciations result in river terrace formations (*below*).

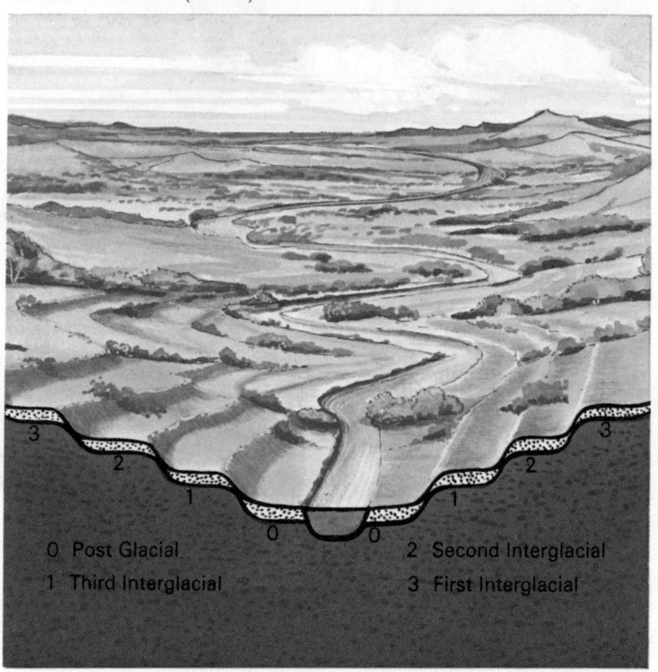

Modern extent of the north polar glaciation

Maximum extent of the north polar ice in the Würm glaciation

The glaciations

Glaciations are known to have occurred long before the Pleistocene Period and on at least two previous occasions polar ice caps formed and glaciers spread towards the equator. Why should these cold periods occur? Several theories have been put forward such as variations in the strength of solar radiation, sunspot activity causing increased snowfall and rainfall, or changes in the positions of the poles, which would alter the seasons and have effects on the weather. The present polar ice caps seem to be large but their fate over the next ten to twenty thousand years is unknown. Neither do we know whether the Ice Age is over. It may be that this is simply another Interglacial phase prior to the spread of a new ice cap over northern Europe and America.

Pleistocene glacial and pluvial correlations

	European Glaciations	East African Pluvials
Upper Pleistocene	Würm	Gamblian
Middle Pleistocene	Riss Mindel	Kanjeran Kamasian
Lower Pleistocene	Günz	Kageran

As warmer weather spread after the last glaciation, forests began to appear in northern Europe and North America. In Africa there is evidence of alternating wet and dry phases (pluvials and interpluvials), from the levels of the former lakes and from over Pleistocene deposits. These wet and dry periods may correspond with the glacial and interglacial phases in Europe. In addition, volcanic activity was taking place in Africa particularly along a line of weakness in the earth's crust, that is now known as the Great Rift Valley. This valley extends from north to south through the east side of the continent. At the time of its formation volcanic eruptions threw up clouds of ash and streams of lava that spread widely, engulfing lakes and valleys. Thick layers of Pleistocene deposits in East Africa are also of volcanic origin, such as those at Olduvai Gorge.

Land mammals

The uncertainty of the climate during the Pleistocene Period greatly affected the land mammals, including early man. Many of the modern mammals first appeared during this time but their distribution throughout the world was very different from that of today. Some of the older mammals survived well into the Pleistocene, such as the sabre-toothed tiger (*Machairodus*) and a larger elephant-like animal that was known as *Deinotherium*.

Another interesting aspect of the Pleistocene fauna was the occurrence of many giant forms both in Africa and in Europe. From the deposits of East Africa there have been recovered the bones of giant bovines (*Pelorovis*), giant pig (*Afrochoerus nicoli*), giant hippopotamus (*Hippotamus gorgops*) and a baboon of terrifying proportions, which has been given the name of *Simopithecus*. Similarly from Europe, in rather later deposits, the bones and antlers of a giant deer, known as *Megaceros giganteus*, have been discovered.

Machairodus, the sabre-toothed tiger

Deinotherium

Glacial and interglacial periods

As the glaciations advanced and retreated, the fauna of Pleistocene Europe alternated between those groups of animals best suited to either cold or warm/temperate conditions.

During the last glaciation, mammoth (*Mammuthus*), woolly rhinoceros (*Coelodonta antiquitatis*), red deer (*Cervus elaphus*), reindeer (*Rangifer tarandus*), wolf (*Canis lupus*) and ibex (*Capra ibex*) ranged widely over Europe, together with the arctic fox (*Alopex lagopus*), cave bear (*Ursus spelaeus*), bison (*Bison priscus*) and wild horse (*Equus*).

A possible Pleistocene glacial scene (*below*) and interglacial scene (*right*)

Similarly during the warmer interglacial phases animals that would now be regarded as tropical creatures spread away from the equatorial regions and penetrated many parts of Europe. Forms such as the lion (*Panthera leo*), straight-tusked elephant (*Palaeoloxodon antiquus*), fallow deer (*Dama clactoniana*) and Merck's rhinoceros (*Didermocerus merki*), as well as many other tropical animals, occupied northern Europe. Hippopotami are known to have played in the river Thames during the Third Interglacial. The horse (*Equus*), as well as the auroch (*Bos primigenius*), a large bovid from which modern cows and buffalo are derived, roamed Britain during the Würm 1/2 interstadial, a temperate interval.

FOSSILS – THE DATING OF MAN

Fossil bones

It is difficult to discover the type of animal or man that was alive thousands of years ago but the answer lies chiefly in the study of fossil bones. When an animal dies its flesh will be eaten by scavengers and its bones picked clean by birds but what will happen next depends upon the climate and the local soil conditions.

Bones that are swept into streams and rivers become buried in gravel and silt. Excavation of gravel beds and river deposits may bring to light bones from a wide range of species, which

Burial of bones often preserved ancient animal life

were alive at the time of deposition. In this way the fauna of a region can be reconstructed.

In some volcanic regions eruptions will produce lava flows and dense clouds of wind-blown ash. Animals trapped by the choking dust may become buried rapidly and, therefore, well preserved. Ash may settle on land or on still water surrounding volcanic zones but in either case preservation conditions are good. Changes in the level of lakes may cover bones over a wide area near the waters' edge. Some of man's early ancestors were trapped in this way at Olduvai Gorge in East Africa.

Cave deposits

Deposits in caves are generally good hunting-grounds for fossil bones and caves most commonly occur in limestone regions since percolating waters dissolve the alkaline rocks and form underground rivers and large caverns. Similarly the sea will attack limestone cliffs to form a type of karstic cave. Many animals, including man, seek shelter in caves and the remains of their food become buried in the cave floor. Roof

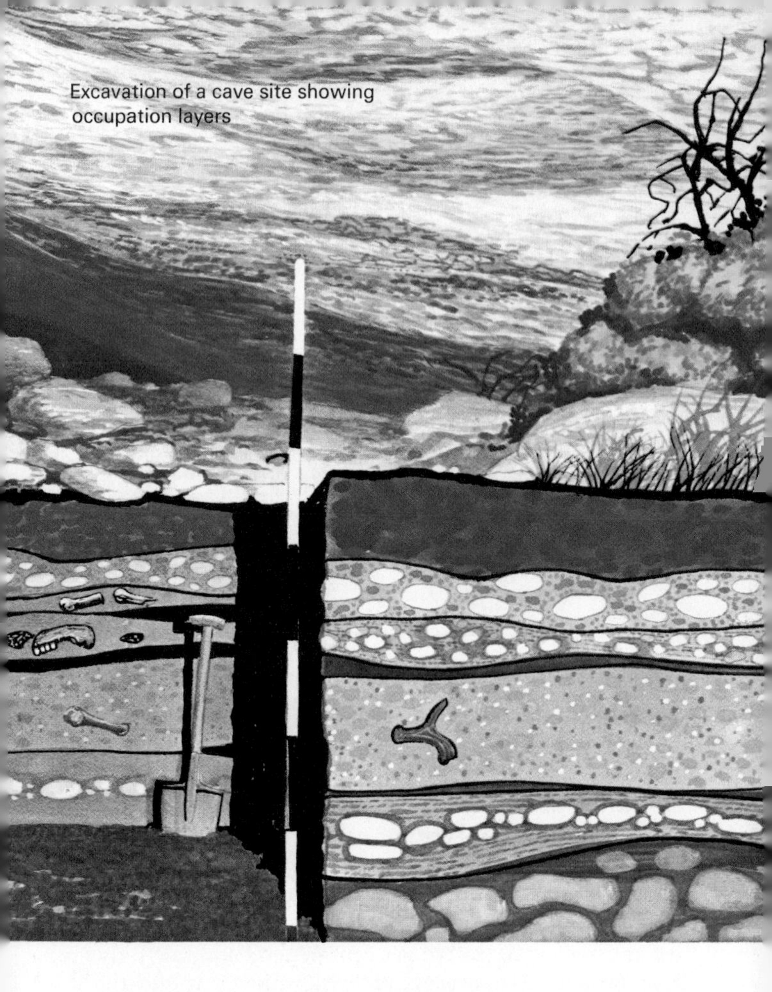

Excavation of a cave site showing occupation layers

falls will gradually cover the floor and bury the bones in an alkaline soil which provides a good preservation medium. Caves occupied by early man are often filled with stratified deposits showing well-marked occupation floors. Provided that the layers are undisturbed, this gives good evidence of early man and can often be used to show a sequence of events. Early occupation floors will be near to the bottom while the later floors will be nearer the top.

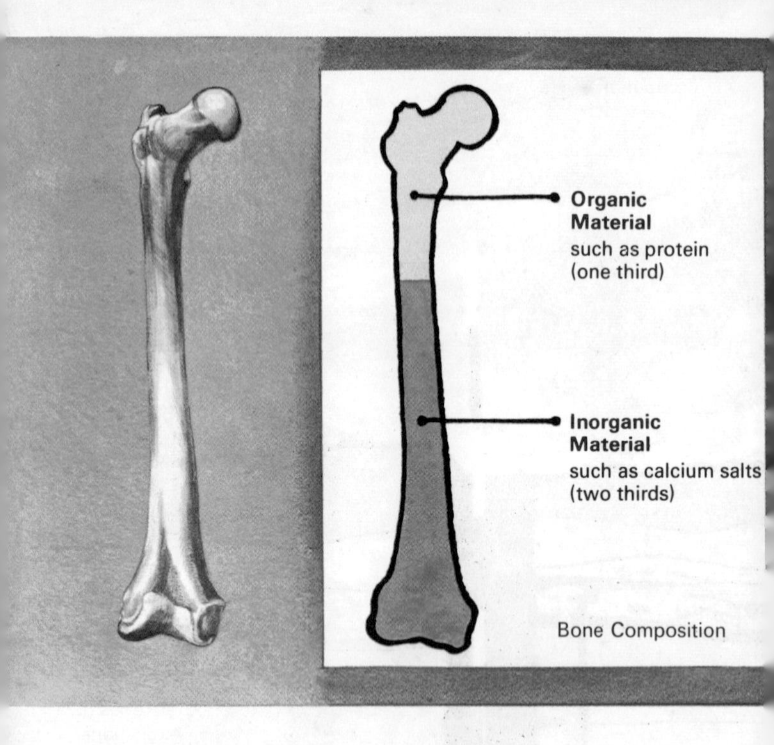

Bone Composition

Bone fossilization

Fresh bone consists of organic protein fibres, known as ossein, set in a mineralized binding substance. Chemically, bone is the same for different animals and contains salts of calcium, magnesium and sodium, in the form of hydroxyapatite. By weight bone is sixty-five per cent mineral and thirty-five per cent protein. The process of fossilization involves the disappearance of the protein material, which is replaced by percolated material from the ground waters in the soil in which it lies. This is followed by a molecular substitution in the crystalline lattice of apatite, which transforms the bone into a mineralized replica known as a fossil. Since the substitutions are at molecular level, the form of the bone is perfectly preserved while it is changing to stone. Soil conditions are, of course, extremely important as the chart opposite illustrates.

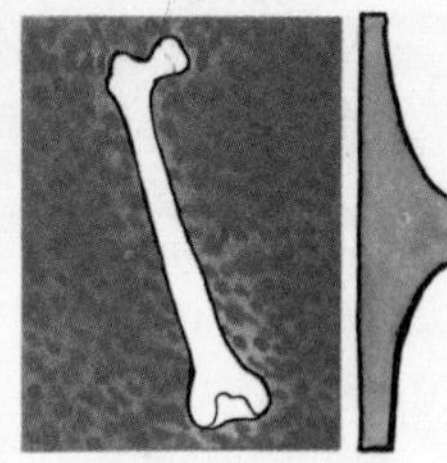
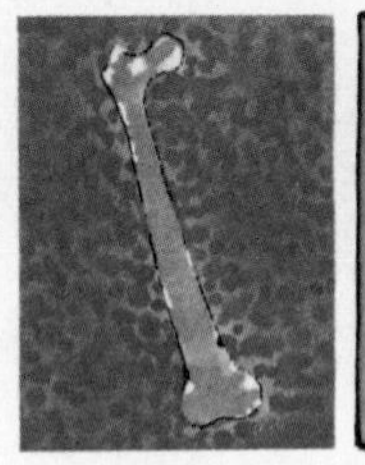

1 Wet Acid Soil

Disappears

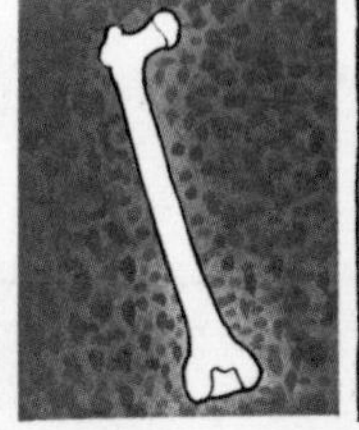
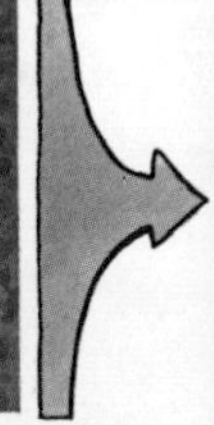
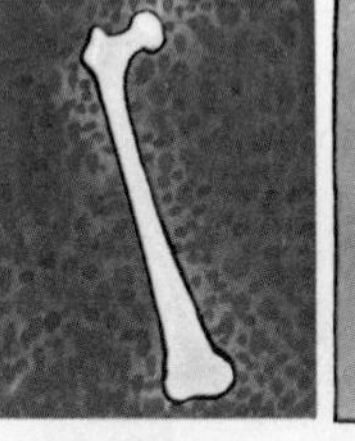
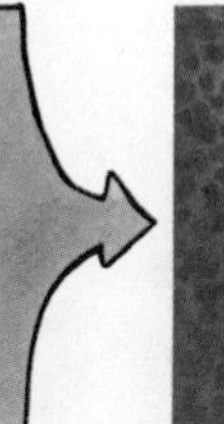
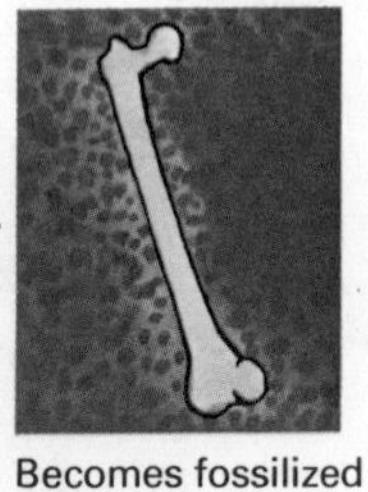

2 Wet Alkaline Soil

Becomes fossilized (heavy)

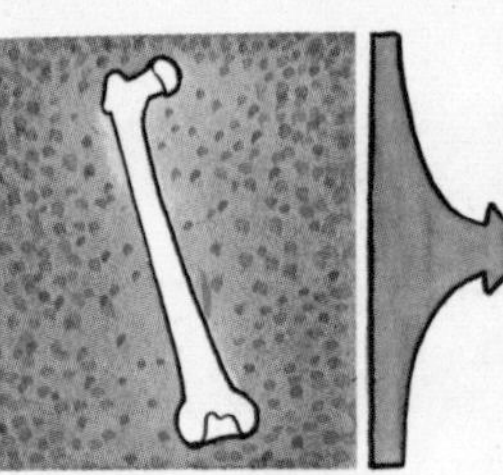
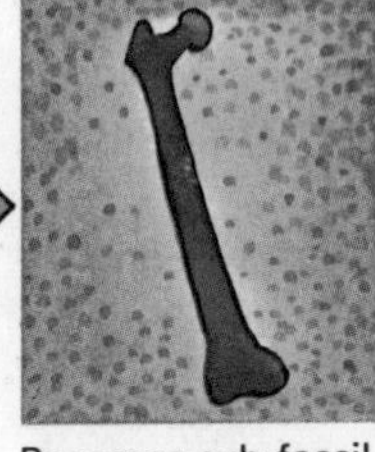

3 Dry Alkaline Soil

Becomes sub-fossil (light)

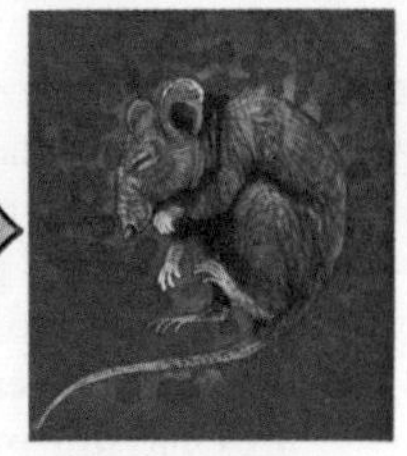

4 Wet Airless Acid Soil (Peat)

Soft tissue and bone preservation

Fossilization and Mineralization

In wet acid soils, fossilization does not occur and bone remains disappear 1 Wet alkaline soils do not dissolve the mineral elements in a bone and fossilization is possible. 2 In a dry alkaline soil, a light sub-fossil is produced 3 and in airless wet acid soils such as peat, complete preservation of soft tissues and bone will occur 4.

Discovery and excavation

The remains of early man are among the rarest and most highly prized fossils of all. Palaeontologists, or fossil hunters, search through likely deposits with infinite patience and care. At a site where it is thought that human remains may be found, such as a limestone cave, the recent soil is removed using hand tools and small trowels; the area having previously been marked out. When a layer is uncovered, every fossil bone and every stone tool is recorded and marked as it is recovered. Should a human bone be discovered it would be cleaned using dental

Excavation on the site of fossil remains

picks and camel hair brushes, in order to preserve every scrap. After this, photographs are taken before the find is removed from the deposit to establish, beyond doubt, that it truly came from this site and has not been placed there by other means.

To remove the bone, each part is covered with wet tissue strengthened with plaster-of-Paris bandages. Then the whole find may be undercut and lifted out in one piece with final cleaning taking place at the laboratory, or, if the fossils are embedded in solid rock, drills, and even dynamite may be used. These risky procedures, however, are rarely needed.

Laboratory work

In the palaeontological laboratory, the plaster-of-Paris casing is removed and the long job of developing the fossil from the matrix can begin. Sometimes months of tedious scraping are needed before the bone is clean. Fossils that are embedded in calcareous material can be developed by immersing them in weak acid, which softens the surface layer that can then be cleaned away. Exposed parts of the fossil must be coated with shellac before immersion or the acid will attack the fossil, itself impregnated with calcium salts. Once the bone is free, repairs may be needed. Broken parts are carefully stuck together at points where contact can be demonstrated. At this stage studio photographs are taken to provide a record of the appearance of the bone and the task of producing working casts in plaster can begin.

Developing a fossil from its matrix and plaster casting (*right*)

Casting fossil bones in plaster-of-Paris, plastic materials or fibreglass and resin, is a skilled technique. One method requires a moulding box or plaster boat fillėd with modelling clay in which the fossil bone is carefully embedded halfway. The exposed surface is coated with a separating compound and a second moulding box applied. Silicone-rubber is poured through a hole and when this has set, the moulds are parted and the clay stripped away. Leaving the fossil in the rubber, the box is inverted and pouring from the other side will produce a second impression. The moulding box is then opened and the fossil removed; when the box is closed again and plaster poured, a cast of the fossil is obtained. Modern plastic materials can produce casts of high fidelity, in fact the plastic can be 'filled', so that the reproduction even has the same weight as the original.

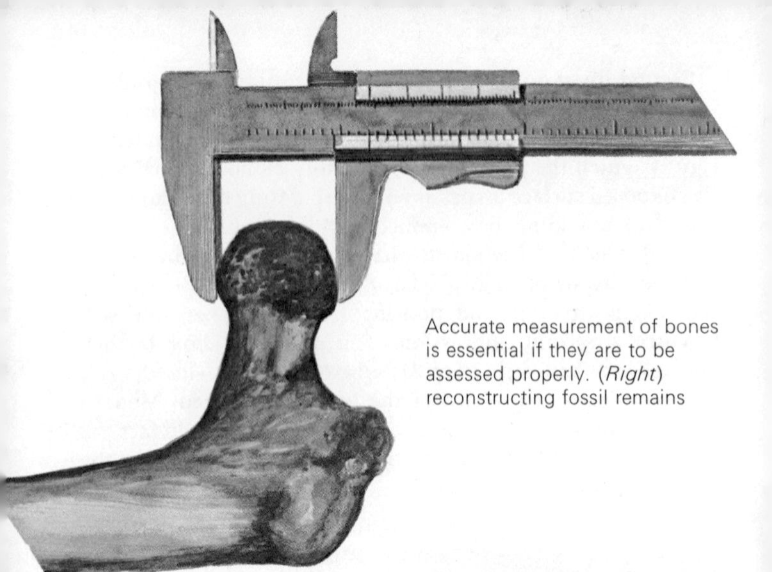

Accurate measurement of bones is essential if they are to be assessed properly. (*Right*) reconstructing fossil remains

Identification

In the anatomical laboratory the tasks of identification and comparison can begin. Anatomists examine the bones and take note of their features and peculiarities.

The problems are complicated for two main reasons:—

(i) There is no guarantee that the group of bones all belong to the same individual or even the same species.

(ii) Fossil finds are often broken and very commonly incomplete and distorted.

However, there are some guiding principles that are of great assistance. First of all higher animals are bilaterally symmetrical, so that missing parts of bones can often be restored by reference to the opposite side. Secondly, it is not difficult to decide to which side a bone belongs, and thirdly it is usually possibly to say whether a bone belonged to an infant, an adolescent or an adult. In this way, a given assemblage of bones can be ascribed to a minimum number of individuals. At this point bones that are believed to belong together are grouped, and those that form a functional complex such as a hand, a foot, a pelvis, or a skull will be articulated. In this way skeletons are built up from the hotchpotch that has come from the deposit.

It is rather like doing a three-dimensional jig-saw puzzle that has half the pieces missing and the rest broken.

To assess each bone properly it is necessary to measure it accurately. To do this, anatomists use vernier calipers which are marked in millimetres and also dividers, calibrated steel rules and protractors. In this way new specimens can be compared with others of the same kind, and with comparative series of human and primate bones.

Statistical analysis

No two people are *exactly* alike in their body build and appearance. Normal variation between individual members of a species is a biological rule based on the varied genetic background of each member. We inherit a kaleidoscope of characters from our parents and grandparents, which ensures that we have our own unique characteristics. Moreover, normal variability is not only a feature of external appearances but it also applies to internal structures such as muscles and bones. Since people differ both in size and shape, it follows that a population must consist of a number of individuals within roughly definable ranges of size and shape. Within these limits it is likely that most will be about average, while a few will be very large and a few very small. Factors such as size and shape will be distributed about a mean value, the average for that population.

Fossil bones are random samples, often small in number, representing an unknown population. Therefore statistical

Use of computers permits extensive statistical analysis.

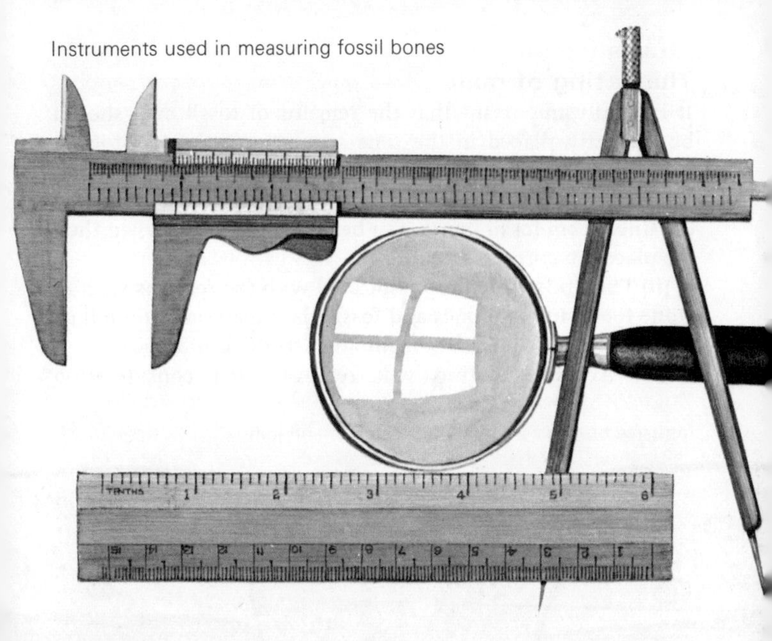

Instruments used in measuring fossil bones

methods in palaeoanthropology are commonly directed towards comparing single fossils or small samples with larger samples from populations of related species.

In the past, statistical analyses have been of a comparatively simple nature and have helped anatomists in the interpretation of the information that can be obtained from simple examination of the bones. The stoutness of the shaft and the size of the muscle markings on a leg bone will indicate the power of that limb and the shape of the surface of a joint will tell how that joint moved. If several joints in a limb are available, the way that limb functioned may be deduced.

The advent of the computer has added new and exciting possibilities by permitting extensive statistical calculations to be completed in seconds. Using the techniques of multivariate analysis, it is becoming possible to group fossil specimens and to estimate the distances separating the groups. In association with assessments of the structure and function of fossil bones, these distance measurements can be a help in the great task of classifying man's ancestors.

The dating of man

It is vitally important that the remains of fossil man should be correctly placed in the time sequence of the Pleistocene Period for three reasons:

(i) The evidence of evolutionary progress, which can be obtained from fossil bones, can best be interpreted when they are placed in correct order of age.

(ii) The finds which are associated with the remains such as stone tools, animal bones and fossil plant materials will tell us about the climate, environment and activities of fossil man.

(iii) Accurate dating will remove from consideration

Intrusive burials can give a false idea of antiquity.

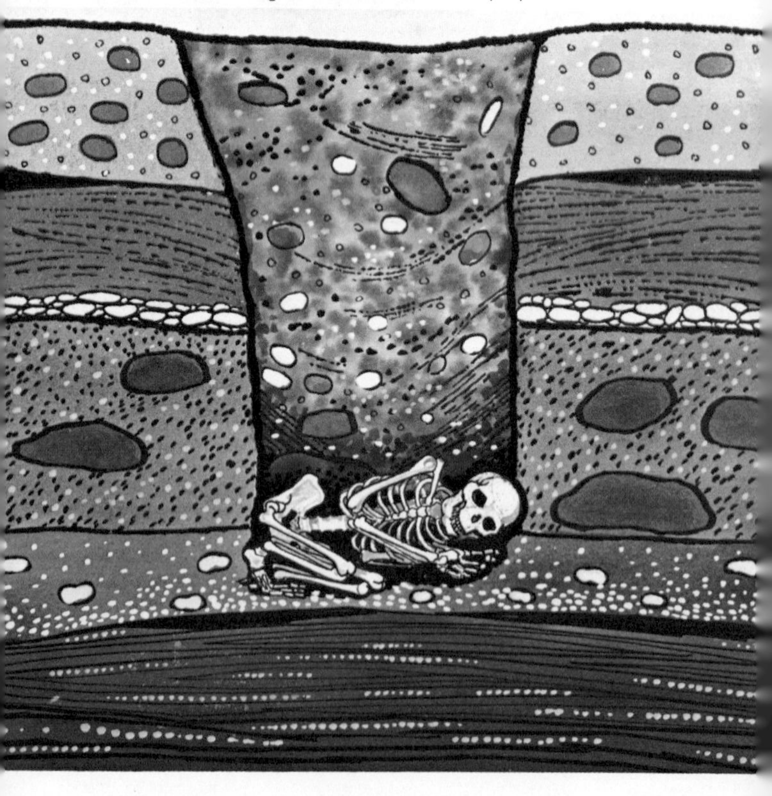

any fraudulent 'specimens' and genuine mistakes.

Rocks are laid in sequence, the younger rocks overlying older rocks, but intrusive burials into ancient deposits can mislead the unwary into giving a skeleton an antiquity that it does not deserve. A new find must, therefore, be thoroughly checked to ensure that it belongs with the deposit in which it was found. It is also important that the deposit is examined to make certain that it is undisturbed and does not consist of a mixture of materials redeposited from another site. Chemical methods are used to establish the contemporaneity of a specimen and its deposit.

Fluorine tends to accumulate in buried bone, its rate of accumulation depending upon the concentration of this element in the surrounding ground waters. Clearly this rate will vary from time to time and from place to place so that any estimate of absolute age from fluorine concentration must be uncertain. However, if there are a number of bones in the deposit, animal or human, which have been there for the same length of time, they should have about the same fluorine concentrations. In this way human bones may be shown to be contemporaneous with those of extinct animals, such as the mammoth or the sabre-tooths.

Similarly, the nitrogen content of bone tends to diminish when the bone is buried and equivalent nitrogen contents can also be measured and suggest the contemporaneity of two bones from the same site.

When the famous Piltdown fossil man was first suspected of being a fraud, and perhaps the result of a hoax, chemical analyses were performed on the individual bones of the find, and on other animal bones which were recovered from the same deposit. The results of these analyses illustrated quite clearly that these finds did not all belong together, since their chemical compositions did not match. This proved that these bones could not have all been in the same deposit for the same length of time, and, therefore, that they were not all contemporaneous.

Ever since this exposure of the Piltdown material as fraudlent, all new specimens are rigorously analysed. This method thus ensures that, with future finds, there shall be no repetition of this misleading situation.

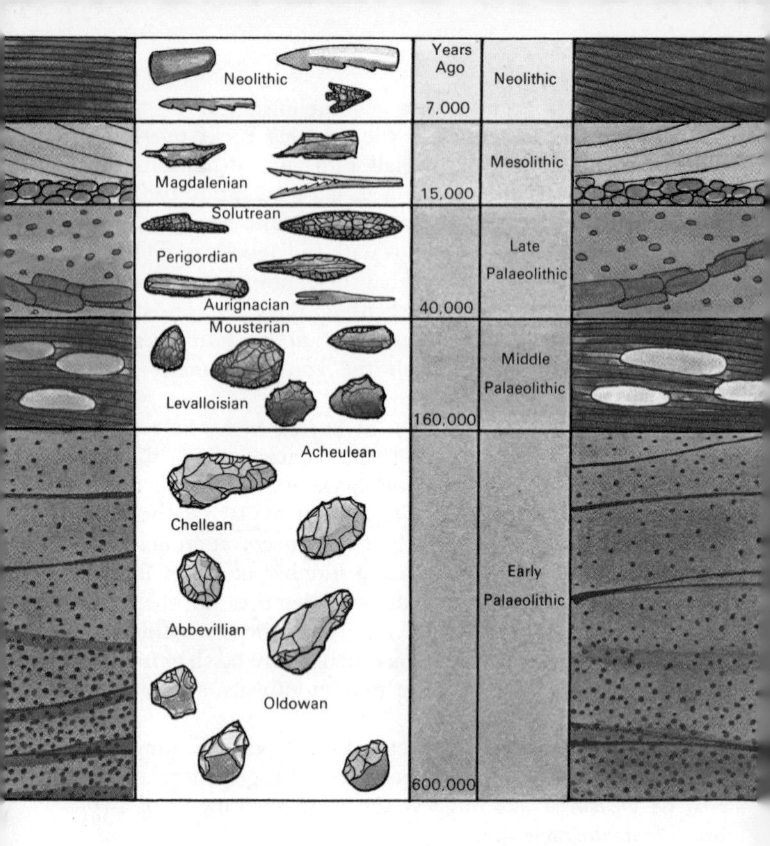

Stone tools in stratified deposits aid the dating of fossils.

Geographical and botanical analysis

After establishing that a fossil bone existed at the same time as the deposit in which it was found, new evidence which is derived from the deposit can be considered. Examination of the local rocks (geology) and the layering of the deposit (stratigraphy) will help in establishing the age of the fossil in relation to the local sequence. In addition, examination of the remains of animals from the same layer as well as the stone tools found on the site, which is known as faunal and archaeological dating, will contribute to the assessment of relative age by permitting comparison with known sequences of finds from other sites. In

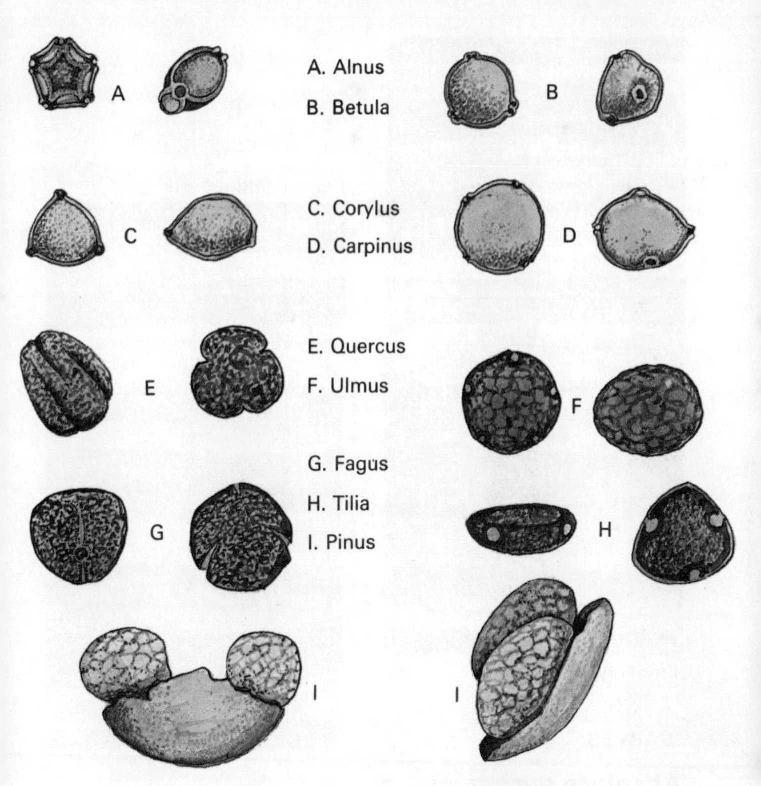

Pollen grains of post-glacial trees (*above*)

general terms, palaeontologists are familiar with the succession of animals throughout the Pleistocene Period in the same way that prehistorians can identify the succession of stone tool cultures.

Botanical specialists are called palynologists and are able to identify pollen grains that are obtained from the various strata of a deposit. In this way they can determine the vegetation and climate which corresponded with each layer. When a pollen 'profile' has been constructed, it can be compared with those from other sites and the relative dating of the deposit may be confirmed or refuted.

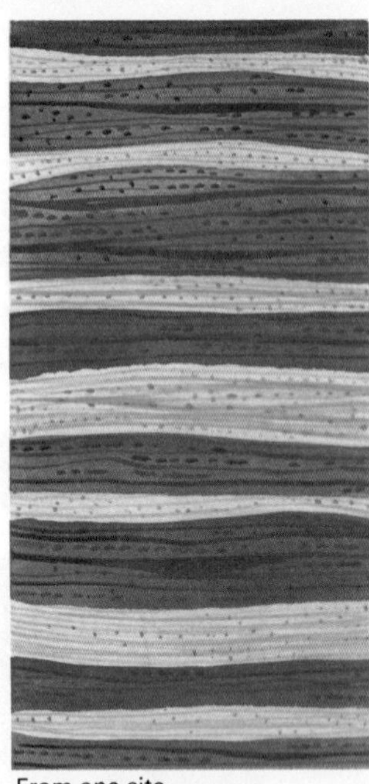

From one site

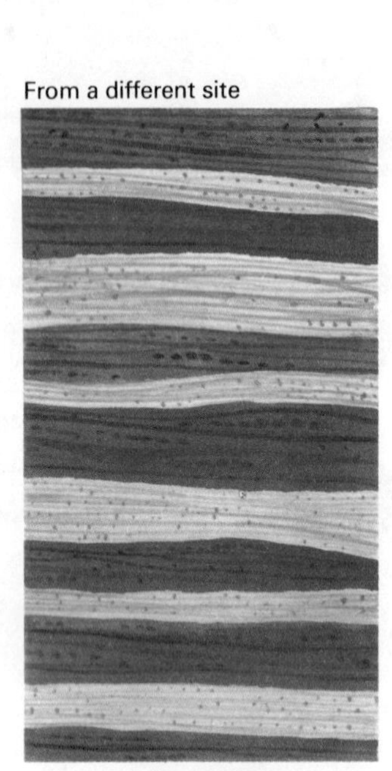

From a different site

VARVES

Absolute dating

The final step in the dating procedure must be an attempt to establish the age in years, before the present, that the fossil man was alive. This is known as the *Absolute* or *Chronometric Dating* of the fossil. Recent advances in the science of geochronology have added a number of new methods which are greatly increasing the span of years over which absolute dates may be obtained. The reliability of these dates depends upon the accuracy of the method and upon the material used for the investigation. Comparatively few absolute methods can be performed on fossils themselves, either because they are not suitable for the technique or because their use would involve their destruction. If the fossil is not to be used, then great care must be taken to ensure the contemporaneity of the material which is being measured and the fossil itself.

One of the oldest methods of absolute dating is based on

Tree Ring Counting

varve counting. It was observed at the end of the last century that glacial clays in parts of Sweden are laminated, the layers occurring in pairs, or varves, and representing the sediments left by the yearly melting of the ice towards the end of the last glaciation. In a few places varve deposition has continued up to the present, thus it has been possible to establish an absolute chronology in northern Europe which goes back 20,000 years to the end of the Pleistocene Period.

Similarly, tree-ring counting, or dendrochronology, has permitted comparison and correlation of ring patterns found on cut surfaces of recent and ancient trees. This method is, however, severely limited by shortages of suitable material.

Radioactive dating

Isotopic dating is the most widely used of the absolute dating methods. All the processes depend on the fact that naturally-occurring radioactive isotopes gradually decay with the passage of time. As the decay rate for each isotope is known, it is possible to calculate for a given specimen the length of time during which the process has been taking place.

Perhaps the most important direct method of isotopic dating is the radiocarbon estimation first applied in 1959. This method has provided a chronology which extends back for about 70,000 years. Nitrogen atoms in the upper atmosphere are bombarded by neutrons produced by cosmic radiation resulting in the production of a known proportion of radioactive carbon (C^{14}), that becomes incorporated in atmospheric carbon dioxide. In turn, this carbon dioxide is absorbed by vegetation and passes into animal tissues when the plants are eaten. When the animal dies no further isotope is absorbed and

A sample placed in a heating tube is converted into carbon dioxide.

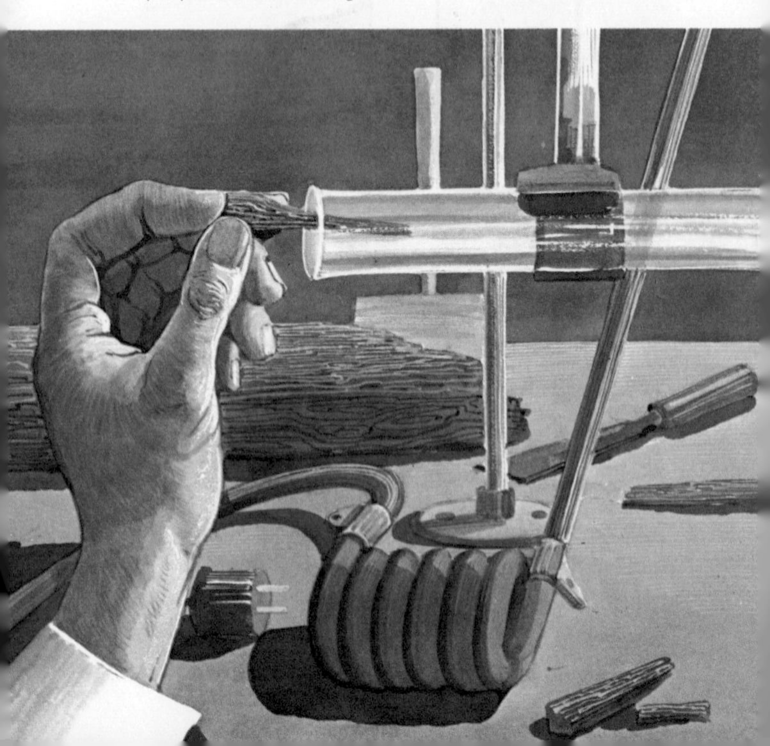

beta ray emission gradually reduces the radioactivity of the remains to about a half after a period of 5,730 years, the 'half-life' of the isotope.

Estimates of the residual radioactivity in a specimen, made with sensitive Geiger counters, allows calculation of its absolute age. It is fortunate that organic carbon is a constituent of all living material, therefore wood, bone, charcoal, peat, horn and vegetable remains in soil can all be examined. On the other hand, the diminution of the radioactivity limits the method to about 40,000 years without the use of special techniques. Above 70,000 years the amount of radioactivity is too small to estimate. On account of the errors, which inevitably arise in investigations of this kind, radiocarbon dates are usually quoted with a range of tolerance, for example, 6,570 ± 340 years. Old carbon, such as that found in Carboniferous coal measures, is radioactively 'dead' and therefore useless for dating by this method.

A Geiger counter measures the residual radioactivity of a specimen.

Potassium argon and uranium

Another radiometric dating method, the potassium-argon technique, depends on the fact that naturally occurring potassium contains an isotope potassium-40 (K^{40}) that decays at a known rate to the inert gas argon-40 (Ar^{40}) which becomes trapped in the crystals of potassic minerals. Estimates of the argon content of a sample of one of these minerals, obtained from a deposit containing fossil bones, will indirectly measure the age of the bones. The minerals which are available are usually derived from volcanic deposits. Fortunately the half-life value of K^{40}, which is approximately 1,300 million years, is considerably greater than that of C^{14}, thus extending the range of radiometric dating. In practice potassium-argon dating is commonly limited to deposits that are greater than 20,000 years old.

The process involves taking a known weight of the sample, carefully selected for its potassium content, and heating it in an electronic furnace to drive off the contained argon. Analysis of the mineral to determine its potassium content, coupled with estimation of the amount of argon produced from the sample, enables geochronologists to calculate the age in years of the specimen. This is achieved by comparison of the experimental result with theoretical results obtained from the known decay rate of potassium to argon.

In order to obtain accurate and consistent results from this method, great care must be exercised in the selection of samples to ensure that the specimen used in the estimation is contemporaneous with the deposit.

A new method of isotope dating has been developed which will date cores of silt drilled from the sea bed. Uranium (U^{234}, U^{238}) in sea water decays to form proactinium (Pa^{231}) and thorium (Th^{230}), whose half-lives are 34,300 and 80,000 years respectively. The new elements are picked up by sediments forming on the bottom and become incorporated in the sea bed.

Biological examination of the constituents of deep sea cores can throw light on past climatic changes by the prevalence of cold or warm water forms within the sample. Before sea cores can be of wider use to assist in the dating of fossil man, however, far more accurate correlations between sea bed changes and land deposits must be attained by examining core samples from many more locations and from a range of oceanic sites.

The pattern of alternations of warm and cold sea water animals and plants should, therefore, providing the deposition of deep sea sediments is constant in rate, reflect the pattern of glacial and interglacial periods in the world's climate. This may work fairly well for the more recent glaciations, however, but can give rise to difficulties with the earlier climatic fluctuations. It is hard to say, for example, whether a given cold phase represents the Günz glaciation or one of the several pre-Günz minor glaciations.

Cores of silt from the sea bed also give indications of past climates either by biological examination, or by a newly developed method of isotope dating based on the decay of uranium.

Core sample from a deep sea bore

Study of fission tracks

The latest method of isotopic dating was developed after it was discovered that uranium contaminents in old glass tend to explode and produce tiny fission tracks. These minute 'atomic explosions' can be made visible to the microscope by etching the glass with suitable acid. The number of these 'explosions' is counted for a given area, then the specimen is irradiated in order to explode all the remaining fissile atoms, and the specimen recounted. Since the rate of fission under natural conditions is known, the difference in the counts will permit calculation of the time between formation of the glass and its examination. It was a short step to apply the method to the naturally occurring glasses, such as obsidian, which are found in volcanic deposits.

This advance is of very great importance since volcanic deposits, that are good sources of fossil material, may be dated by two radiometric methods independent of one another. Each is subject to experimental error, but the errors are quite different in each technique. At Olduvai Gorge, in East Africa, both methods have been applied to the volcanic deposits in

Major Divisions of the Pleistocene and Holocene Period in Years BP

	Holocene	Recent Post Glacial	10,000
Pleistocene	**Upper**	Würm Glaciation Riss-Würm Interglacial Riss Glaciation	100,000
	Middle	Mindel-Riss Interglacial Mindel Glaciation Günz-Mindel Interglacial	450,000
	Lower *Villafranchian*	Günz Glaciation Pre-Günz stages	1,000,000
		Plio-Pleistocene boundary	Approximately 3,000,000 years BP

Bed I and the age estimates have tended to confirm each other.

All the known remains of fossil man have been attributed to deposits of the Pleistocene Period, which began at least one and a half million years BP (Before the Present), and extended to the Recent or Holocene Period. Climatic fluctuations were world-wide during the Pleistocene Period and the glacial and interglacial periods had a profound affect on the landscape. It is during investigations of the rocks and fossils deposited at this time that the various means of isotopic dating have proved so valuable.

Fission tracks in obsidian and natural glasses

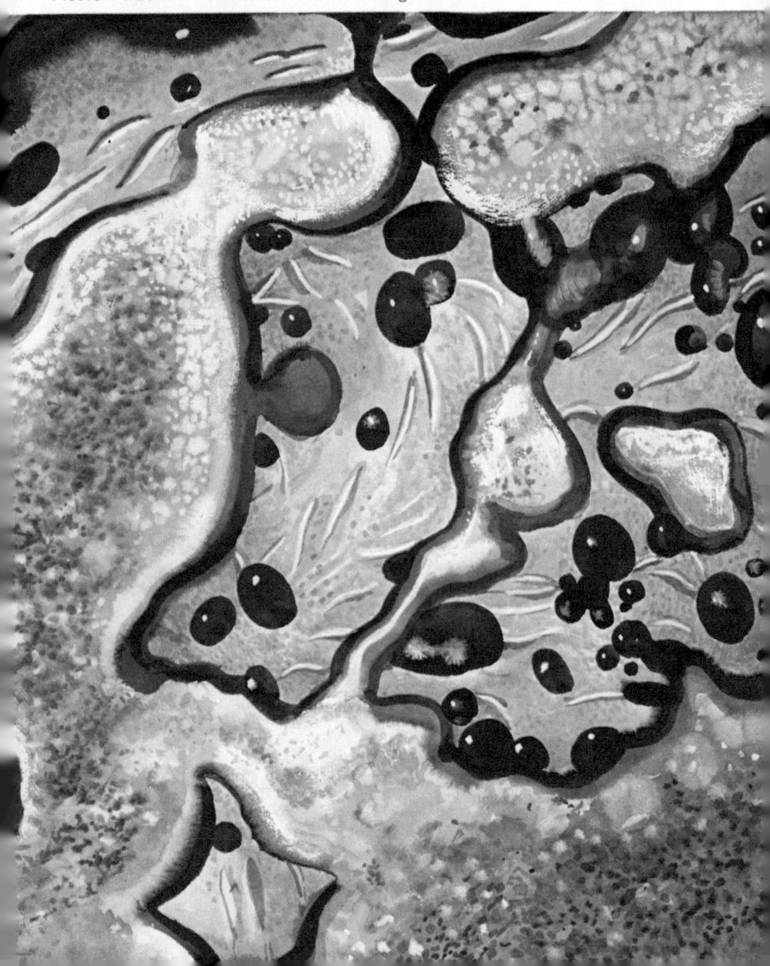

THE STRUCTURE OF MAN

Bone structure

Obviously the remains of fossil men are principally bones. Very rarely are the conditions of preservation such that soft tissue structures can be identified. None the less, soft tissues have their effects on bones, in particular muscular impressions, and grooves or holes for vessels and nerves can be seen. Bones are traditionally symbols of death, as in the skull and cross-bones emblem, but in reality bone is living tissue containing living cells, blood vessels and nerves. If living bone is cut, it bleeds; if it is broken, it heals. Like all living tissue it can respond to stimuli. In particular, bone reacts to mechanical stress in such a way that its shape and internal structure are

X-ray showing stress lines in bone

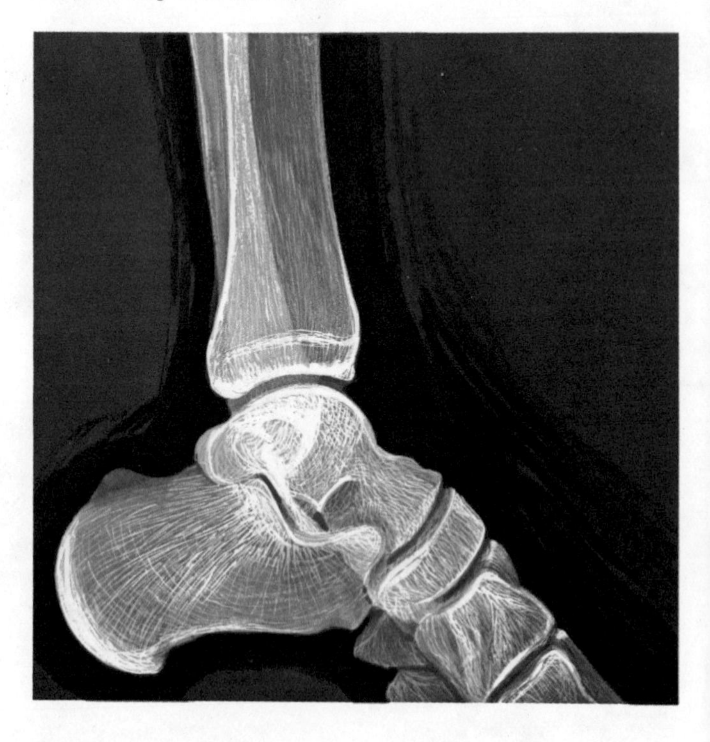

determined by the jobs that it has to do. The strong limb muscles need firm anchorages and the shafts of the limb bones must be able to carry the body weight and transmit the power of the muscles in locomotion. Examination of the bones after death can reveal much of the skeletal function during life and it is on this basis that the anatomical interpretation of fossil bones rests.

Under the microscope compact bone can be seen to consist of concentric plates, or lamellae, arranged around minute canals. In between the lamellae tiny bone cells or osteocytes occupy spaces known as lacunae. These spaces and the larger canals are inter-connected by smaller canals called canaliculi.

Spongy, or cancellous, bone consists of spicules arranged in a lattice which conforms to the principal lines of stress that the bone is called upon to withstand. The spaces between the lattice and the central cavity are filled with marrow. Yellow marrow, which is mostly fat, is present in the shafts of adult long bones while red marrow, or blood-forming tissue, lies inside the flat bones.

Transverse Section of Bone

Longitudinal Section of Bone

Bone composition

Bone is composed of organic and inorganic material and this fact can be illustrated by two experiments. If a bone is burned, it does not char away completely, because a calcined 'skeleton' of the bone will remain, made entirely of inorganic material. It will, however, shatter at a touch. On the other hand, if a bone is immersed in acid the inorganic constituents will dissolve and leave a 'rubbery', flexible bone in which knots can even be tied. It is this combination of constituents which gives bone its rigidity together with its resilience.

Most bones develop from cartilage models laid down in the foetus. Long before birth, bone formation, or ossification, begins near the centre of the shaft. The process spreads towards

Shattering calcified bone

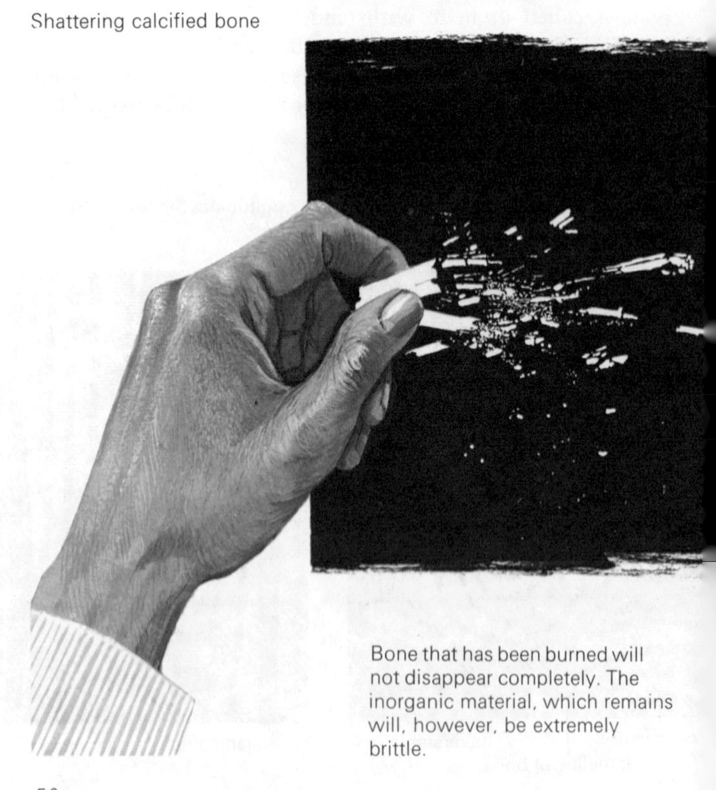

Bone that has been burned will not disappear completely. The inorganic material, which remains will, however, be extremely brittle.

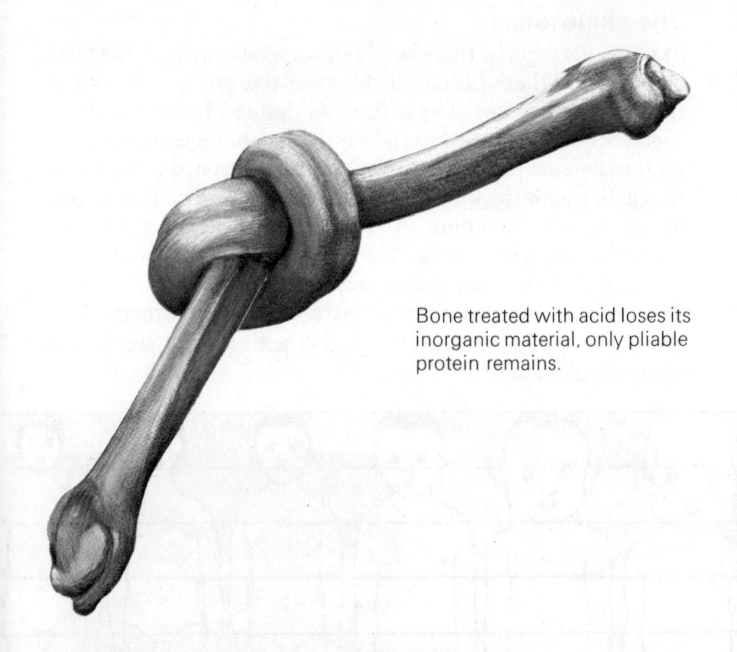

Bone treated with acid loses its inorganic material, only pliable protein remains.

the ends of the bone and then stops and new centres of ossification appear to complete the ends. Growth takes place at the junction between the shaft and the end of the bone and when growth in length is complete, the end fuses on to the shaft. Simultaneously, bone is laid down to increase the thickness of the shaft, and remodelling leads eventually to the adult shape. It can clearly be seen from this that some estimate of age can be obtained from the degree of skeletal maturity of any fossil remains although once again variability is common and this method is not particularly accurate.

Like other parts of the body bones are subject to disease, but they may also reflect the presence of conditions which are more general. Malnutrition and Vitamin D deficiency may lead to rickets, while arthritis of the joints, bone tuberculosis, bone and joint syphilis and inflammation of the bones (osteomyelitis), may all leave their mark. Experts in the study and interpretation of ancient bone pathology are called palaeopathologists.

The skeleton

Man is a vertebrate, therefore his skeleton is based on the same basic plan as that of other members of this group. The axis of the body is composed of a flexible, jointed rod made up of segments of bone called vertebrae. This column is surmounted by the skeleton of the head. The limbs are attached to the trunk by girdles with the lower limb girdle or pelvis firmly attached to the vertebral column for stability, while the upper limb girdle is light and mobile. The upper part of the trunk has a rib-cage that encloses and protects the contents of the chest. The vertebral column, the rib-cage and the skull form the axial skeleton, while the limbs and their girdles form the appendicular skeleton.

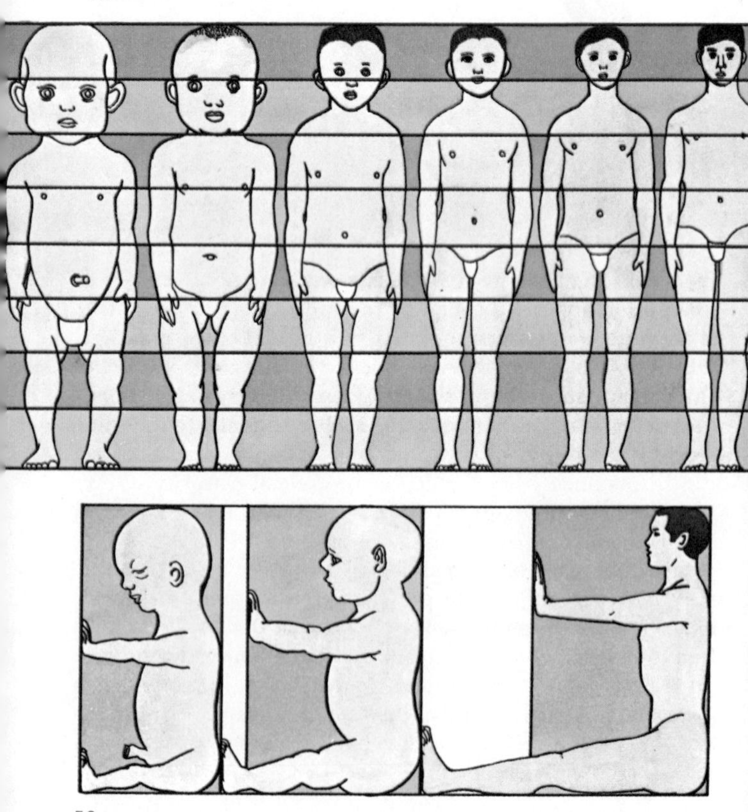

During growth the skeleton not only increases in size, but its proportions also alter. In the uterus, the foetus develops more rapidly at the head end so that the head looks disproportionately large. Principally this is to contain the expanding brain, which grows rapidly and completes its growth comparatively early. Thus, comparisons of head and body proportion throughout the growing period will show an apparent diminution of the size of the head.

Similarly, the limbs of the early foetus develop simultaneously, and are equal in length. As growth proceeds, however, the length of the lower limbs begins to outstrip that of the upper limbs until the adult proportions are reached. The femur which is the longest bone in the body is approximately one quarter of the adult standing height.

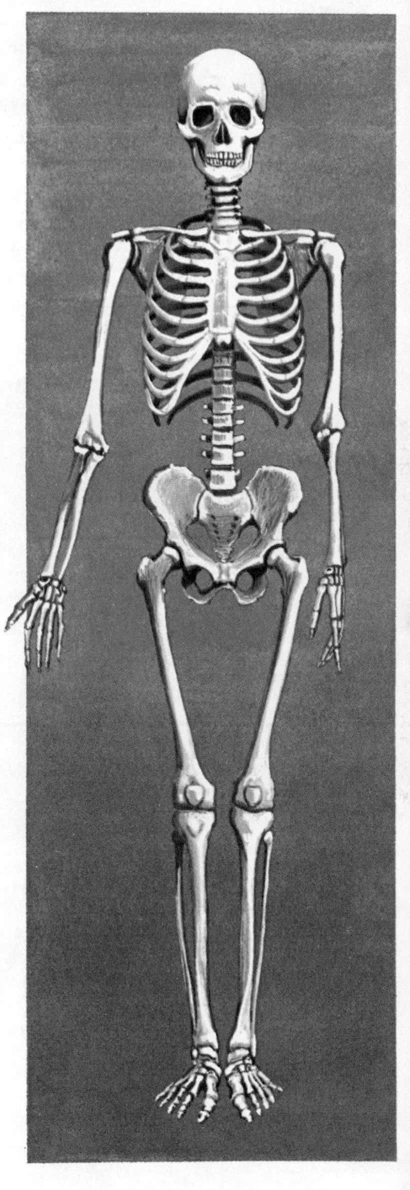

The proportions of the skeleton vary with the age of the individual (*left*) and therefore an infant has a relatively large head in comparison with that of an adult. The skeleton of man (*right*) is based on the same basic plan as that of all the vertebrate animals.

The skull

The human skull is a complex structure made up of two principal parts. One portion, which is called the cranium, forms a bony protective box for the brain, while the remainder acts as the skeleton of the face. The bones of the cranial vault fit together along their edges by means of irregular sutural joints which produce a rigid structural arrangement. Nerves and blood vessels leave and enter the skull through numerous holes and fissures, the largest of which is the foramen magnum in the skull base. Through this hole, the hind brain is continuous with the spinal cord.

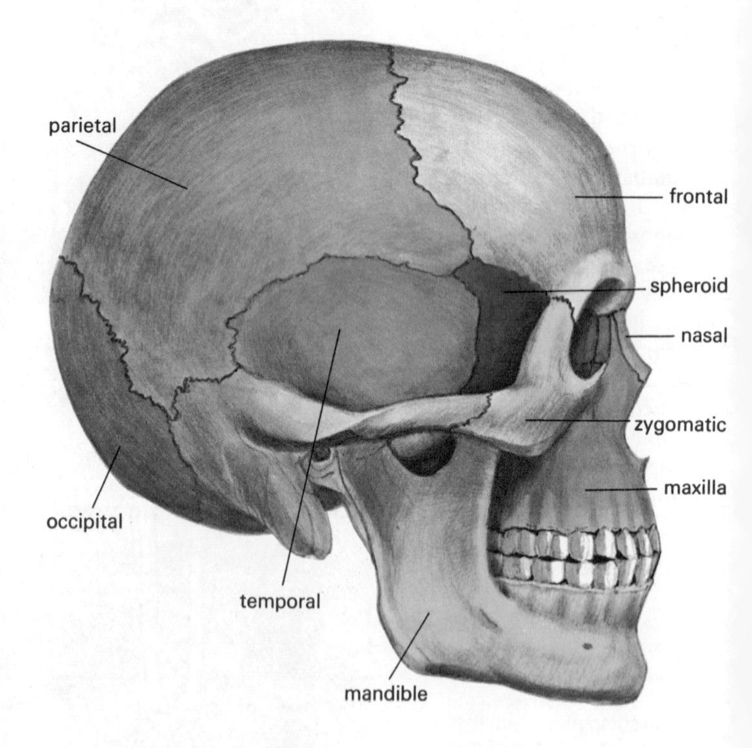

The parts of the skull, which protects the brain and forms the skeleton of the face

On either side of the foramen magnum, the atlas, or first cervical vertebra articulates with a pair of occipital condyles and in this way the skull is balanced on the vertebral column. The facial skeleton is composed of the upper and lower jaws, both of which are slung under the front of the cranium. The upper jaw is firmly fixed and helps to form the bony orbit which houses the eye, while the lower jaw is hinged to permit the movements of mastication. A distinctive feature of the human skull is the lightness of the bones. There is also a well-developed chin, a prominent nose and brow ridges that only form small bumps on the forehead.

Four views of the skull

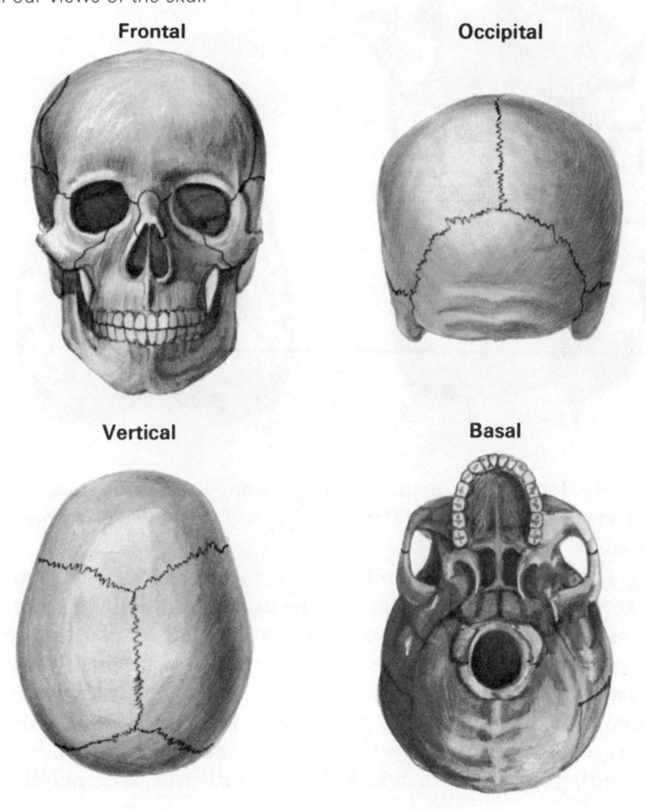

The jawbones

The upper jaw is composed of paired maxillae, which fit together to form the hard palate and the floor of the nose. Within each maxilla is a large air space or sinus which is linked to the nasal cavity. Below this, the teeth are pegged into a stout ridge of bone that is known as the alveolus. The curvature of this ridge and the way in which the teeth are arranged on it are of particular interest, since the rounded, or parabolic as it is called, form of the dental arcade is an important feature

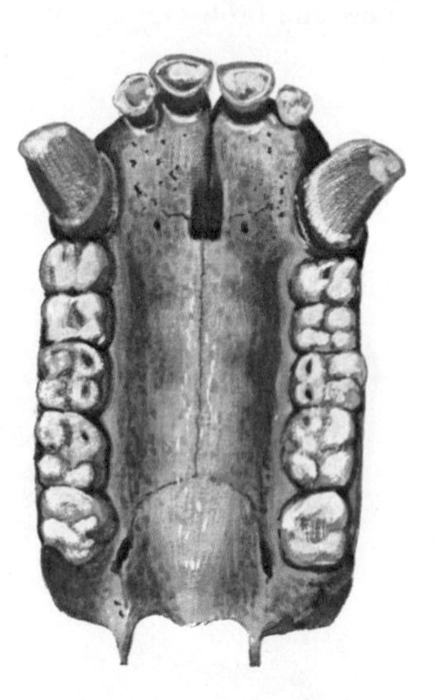

The rectangular dental arcade of apes (*left*) and the U-shaped dental arcade (*bottom*) that is characteristic of man.

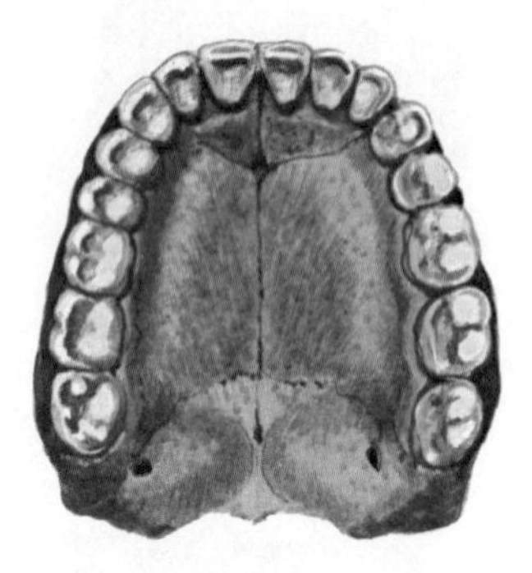

of the dentition of man. By comparison, the dental arcade of modern apes is rectangular in shape and the rows of teeth are parallel.

The lower jaw, or mandible, is a single bone which articulates with the temporal bone on the base of the skull. This joint has a particular form in man since his method of chewing depends upon side to side movement of the mandible. When the mouth is shut the teeth are in occlusion so that the lower dental arcade must match that of the upper jaw. However, the larger back teeth tend to incline inwards to match the corresponding upper teeth which incline outwards. The muscles

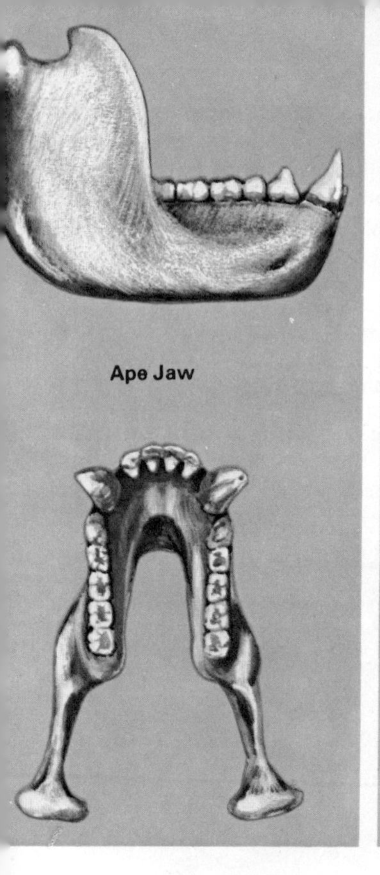

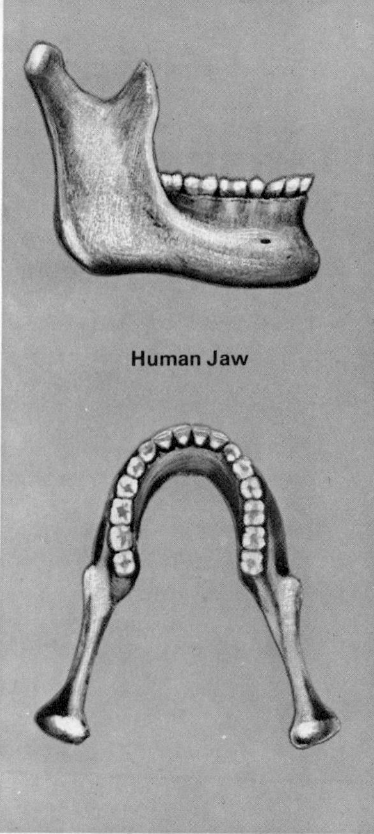

Side view of the jaws of ape and man (*top*) and (*below*) the corresponding occlusal view

which move the lower jaw are powerful and produce well marked processes and impressions on the bone.

The mandible of modern apes is rectangular in shape, chinless and buttressed internally by a ridge of bone that is called the simian shelf. The thickness of the front part of the lower jawbone is on the outside in man and thus gives the characteristic chin lacking in the apes.

Tooth structure

The human dentition is composed of a variety of tooth forms, set symmetrically in the jaws. The deciduous or 'milk' teeth number twenty, while the permanent teeth number thirty-two. The dental formula expresses the distribution of tooth types in the jaws; one half only is represented below:

Deciduous Teeth Incisors (DI) $\frac{2}{2}$ Canines (DC) $\frac{1}{1}$ Molars (DM) $\frac{2}{2}$

Permanent Teeth Incisors (I) $\frac{2}{2}$ Canines (C) $\frac{1}{1}$

Premolars (PM) $\frac{2}{2}$ Molars (M) $\frac{3}{3}$

That part of the tooth above the gum line is the crown of the tooth. The biting surface of the tooth is called its occlusal surface, the remaining surfaces are named buccal, lingual,

Milk and Permanent Dentition

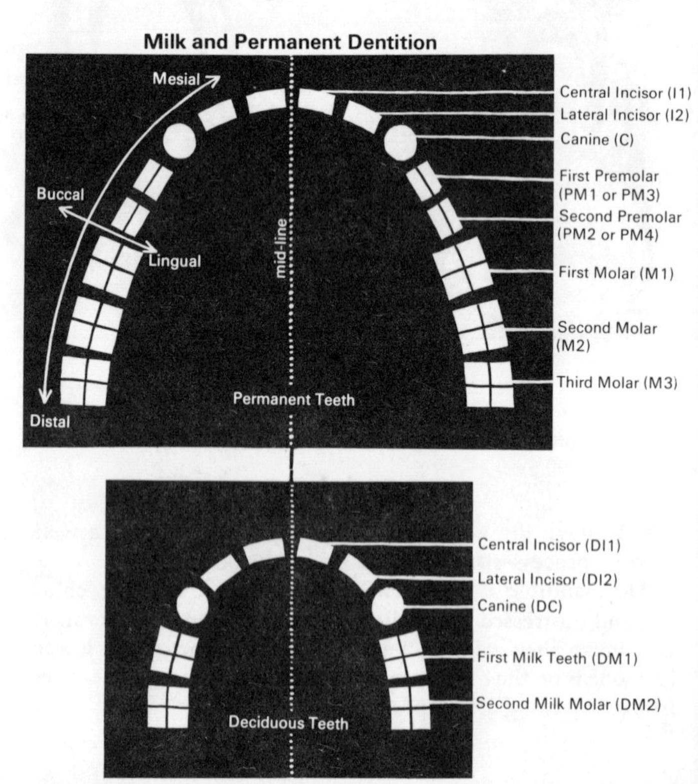

mesial and distal. The occlusal surfaces of premolar and molar teeth carry projections (cusps) separated by grooves or fissures. The human dentition shows reduction in size of the incisors and canines and is also capable of side to side chewing because the canines do not project above the occlusal line. A distinctive flat wear pattern on the biting surfaces is thus formed.

The arrangement of the cusps, however, is basically similar to that of the great apes and resembles that of a Miocene ape known as *Dryopithecus*. This arrangement consists of five cusps separated by a Y-shaped fissure pattern. This cusp and fissure arrangement is often found in the lower molar teeth of fossil man, but in modern man there is commonly a suppression of the fifth cusp and translation of the fissure pattern into a plus (+) form, particularly in the second and third molars.

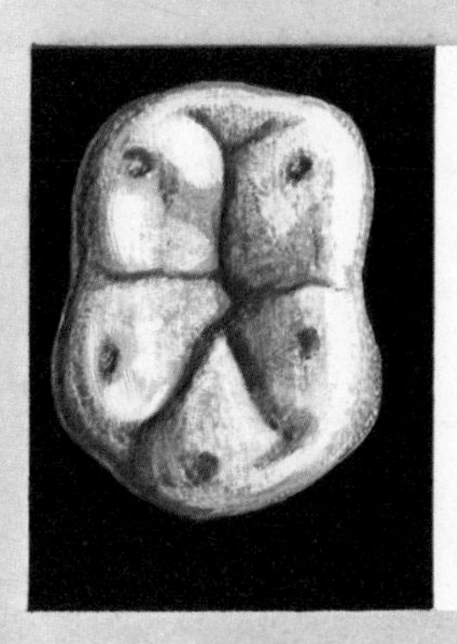

The left lower second permanent molar of a Dryopithicine ape showing a Y5 cusp and fissure pattern.

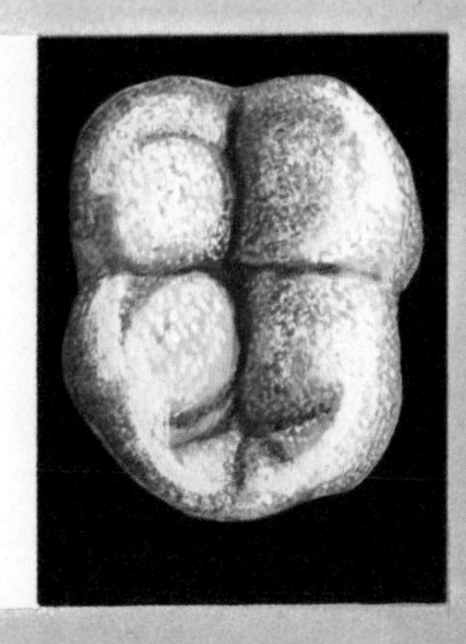

The left lower second permanent molar of a modern man showing suppression of the fifth cusp and translation to the +4 pattern.

Sexual variation

It is possible to sex a skeleton, but, although men are usually bigger than women, and therefore should have bigger bones, this is not always the case. If a mixed population of men and women was taken and sized in a line, nearly all the men would be found at the tall end and nearly all the women at the short end, but in the middle would be a mixed group. There is, thus, an overlap in size.

Sometimes it is possible to sex a fossil skeleton or a single fossil bone because the bones of the pelvis have sexual features. The male pelvis tends to be tall and narrow, while the female pelvis is broad and flat. This results in the male pelvis having a small cavity whereas the female pelvic cavity is large, to make room for the passage of a baby. Sexing single human bones on grounds of size alone is, however, a risky procedure, without a complex statistical analysis.

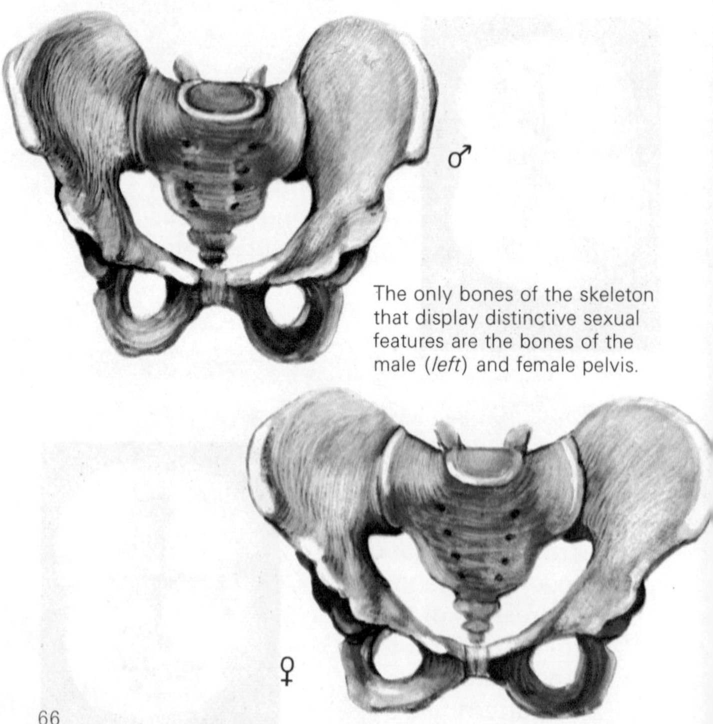

The only bones of the skeleton that display distinctive sexual features are the bones of the male (*left*) and female pelvis.

The vertebral column

The spine or vertebral column is formed by a series of small bones that increase in size downwards and then taper sharply towards the lower end. These bones are joined by discs of fibro-cartilage. The spine appears straight from a front view but has several curves if seen from the side. The foetal spine is concave frontwards, but when an infant lifts its head a cervical curve appears and when it walks the lumbar curve appears. This lumbar lordosis is only fully developed in forms capable of standing upright and walking on two legs.

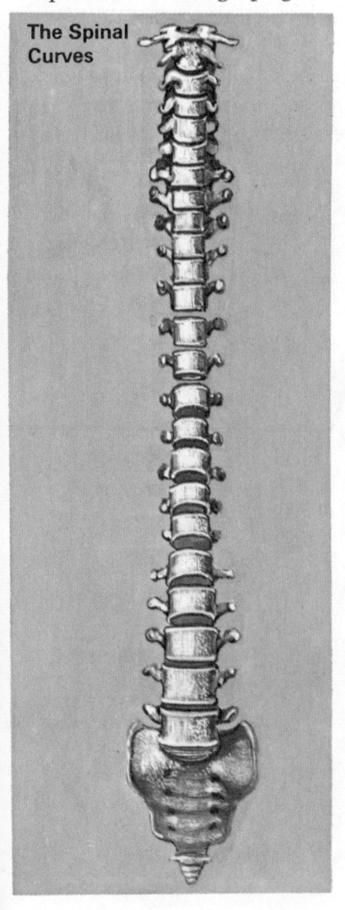

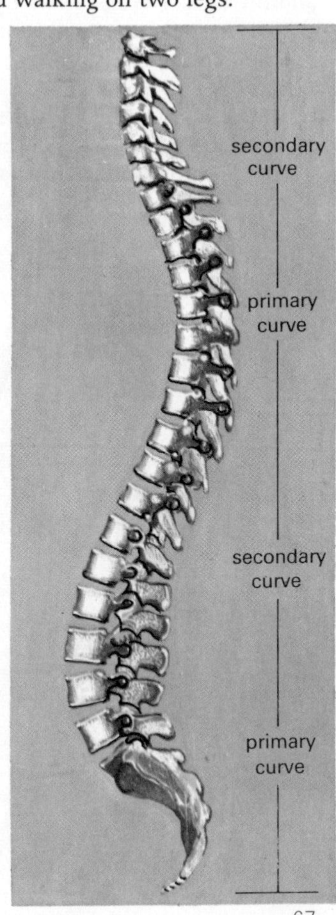

The upper limb

The skeleton of the upper limb consists of a limb girdle by which the limb is attached to the trunk and an upper and lower segment of the limb itself. To the lower segment is attached the skeleton of the hand.

The girdle comprises two bones, the collar bone and the shoulder blade, which are called the clavicle and the scapula. The upper segment of the limb has one bone, the humerus, while the lower segment is formed by two bones, the radius and the ulna. The small bones of the wrist connect the hand to the forearm, and the hand is made up by the bones of the palm, which are called metacarpals, and the bones of fingers, called phalanges.

Movements of the shoulder girdle and the shoulder joint are free, a reminder of man's tree-living origins. Similarly the

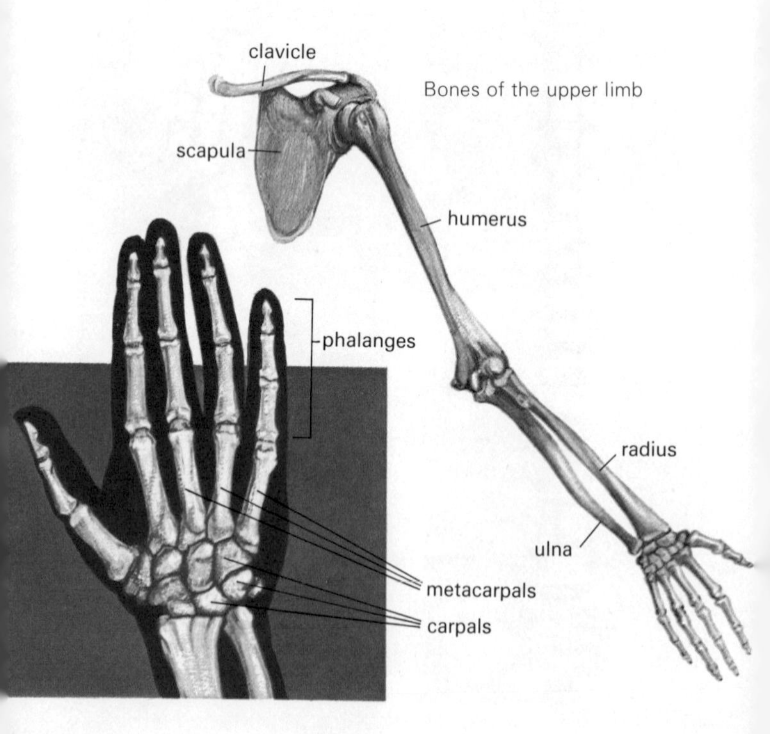

Bones of the upper limb

ability to rotate the forearm and the simplicity of the hand structure also relates to this time. There is little trace of specialized structure in the limb, except perhaps in the range of movement of the thumb and this freedom allows the human thumb and the forefinger to make true pulp-to-pulp contact, true opposition.

Action of the upper limb

The lower limb

The lower limb is attached to the trunk by the pelvic girdle, a robust ring of bone firmly jointed to the vertebral column at the sacro-iliac joint. This is a stable arrangement designed to transfer the weight of the body and the forces of propulsion between the trunk and the legs. The skeleton of the upper segment of the limb is formed by the femur while the bones of the lower segment are the tibia and fibula. The foot is made up of a number of small bones, tarsals, metatarsals and phalanges. In general the upper and lower limbs show clear similarities in their skeletal arrangement but their different functions a reflected in their overall size, proportion, and anatomy.

Movements at the hip joint are fairly free, but the range of these movements does not compare in any way with those of the shoulder. Knee movements are comparable with those of the elbow but rotatory movements between the two bones of the lower segment of the limb are severely restricted. The human foot is a structure that is designed to do two jobs, firstly to support the whole weight of the body and transfer it to the ground, and secondly to transmit the forces of propulsion that are generated by the muscles during walking.

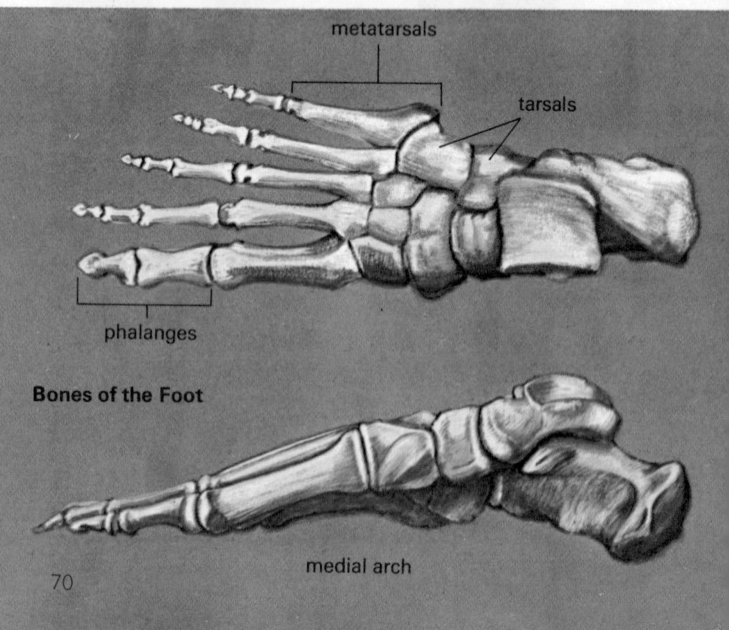

Bones of the Foot

The foot is a curved lever with a short power arm and a longer load arm, the ankle joint acting as a fulcrum. The curvature of the lever is formed by the arched arrangement of the bones, the system of arches being maintained by ligaments, long sling-like calf muscles and short muscles within the sole of the foot.

Action of the lower limb

The hand

The human hand is designed for gripping and is the end product of an evolutionary process that began several hundred million years ago with the development of the pentadactyl limb, which means five toes and five fingers. Other animals have modified this particular pattern by reducing their finger or toe number, like pigs, antelopes, rhinos or horses; but man has retained the simple basic five fingered formula and therefore has an organ of grip which is adaptable to a very wide variety of uses.

In hand gripping, or manual prehensility as this process is called, two basic grip postures can be recognized in man, and these are the power grip and the precision grip. In each of these grips, a different group of muscles and a different side of the hand is involved. The power grip uses the little finger side of the hand while the precision grip uses the thumb side.

The ability to grasp has been of the utmost importance in the evolution of man. The way was opened by the release of the

The power and precision grips of early men (*below*) were developed when upright posture released the fore limb from a weight-bearing role.

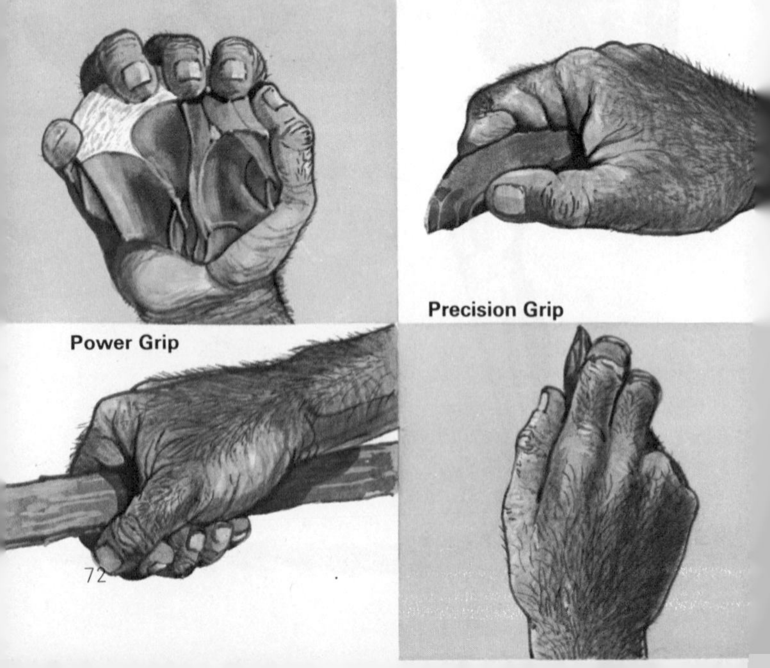

The power and precision grips of modern man

upper limb from a weight-bearing role, when the evolution of upright posture led on to habitual bipedal walking. The hands were, therefore, freed to develop their full potential, for such things as food carrying, infant carrying, weapon carrying and use of various tools.

From the ability to use a stone as a weapon or a crude tool, it is a relatively short step to begin to modify the stone by sharpening its edges. There is some evidence that 'early' human hands were at least capable of a power grip and experiments have shown that it is possible to make primitive stone tools by means of the power grip alone. Later, more sophisticated hand axes and engravers (burins) needed a higher degree of skill in their manufacture and in their use, and this degree of skill demanded a precision grip of almost modern human ability.

Walking

Two-legged walking, or bipedalism, is a complicated and difficult skill. It requires a combination of balance, control and muscular power in the back, hips, legs and feet. To remain upright it is necessary that a line from the centre of gravity should fall within the base. To test this, stand with your feet slightly apart, now lift one foot off the floor. You will sway slightly towards the remaining foot, or fall over. In walking, the foot is lifted and the centre of gravity thrown forward, you begin to fall and while 'falling' the foot that was lifted is put out and a new base for your centre of gravity is provided. The other foot is then lifted and the cycle repeats as you proceed. A line tracing the path of the centre of gravity on the ground will thus have a zig-zag course. Walking can be regarded, therefore, as a controlled fall.

The human foot is a highly specialized organ adapted for bipedal walking on the ground and the sole is almost flat on the floor in the plantigrade position. Human gait can be described as bipedal, plantigrade and propulsive striding. The

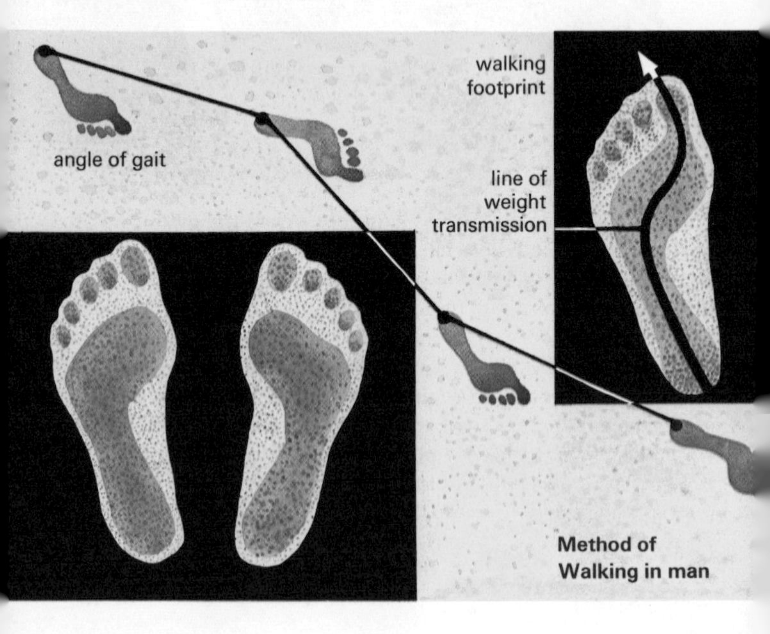

Method of Walking in man

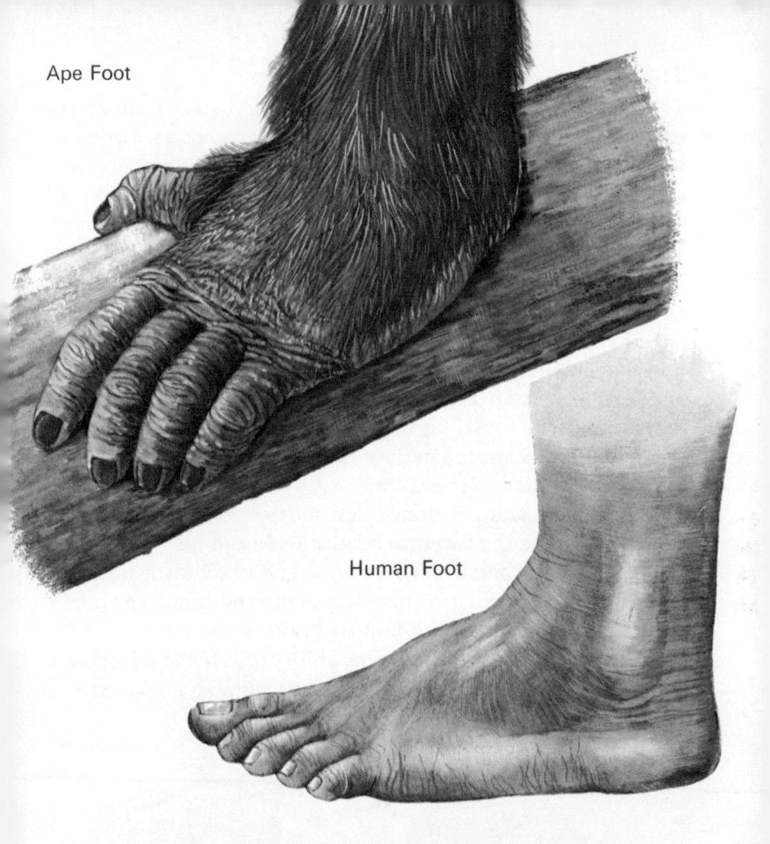

gait of man's nearest living relatives, the modern apes, is very different for the ape foot is designed to grasp and has a divergent great toe. The position of the ape great toe is such that the sole does not lie flat on the ground but adopts an inverted posture. If an ape foot is placed on a curved surface, the divergent great toe 'opposes' the toes and allows a secure hold. Each foot, ape and human, is adapted for the mode of life of its owner and comparisons of the one with the other, although of interest, must not be used to support any far-reaching conclusions, since modern man and modern apes are each at the ends of their own evolutionary lines. The origins of bipedalism must be sought in the foot structure of fossil forms that are within the hominid, and not the ape, line.

The brain

The brain consists of a mass of tissue in the cranium at the upper end of the central nervous system and its functions include the reception and interpretation of information from the surroundings and the formulation and initiation of responses to that information. To this can be added its capacity for memory so that its reactions to present situations may be conditioned by previous experience.

General sensory information is conveyed to the brain by peripheral nerves, originating in sensory receptors in the skin, muscles and joints. Special sensory information reaches the brain from special receptors, such as the eye, the ear, the tongue and the nose, through cranial nerves. Special sense organs are concentrated in the head and are largely responsible for the elaboration of the brain.

The human brain is dominated by two masses of nerv tissue known as the cerebral hemispheres and below them are two smaller cerebellar hemispheres. The brain stem links the cerebral and cerebellar hemispheres with the spinal cord.

In functional terms the human brain is the most complex brain in the animal kingdom. The ability to achieve conceptual thought, to speak, to write, or to master numbers, all testify to

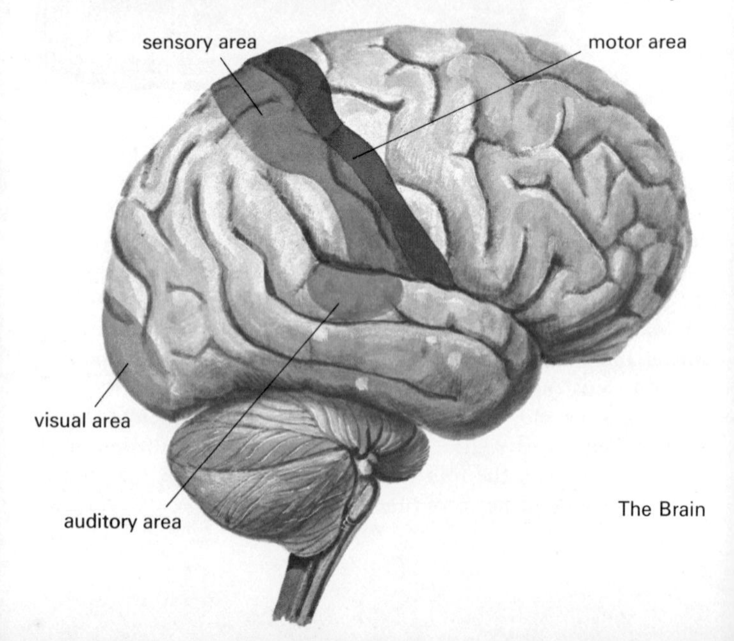

The Brain

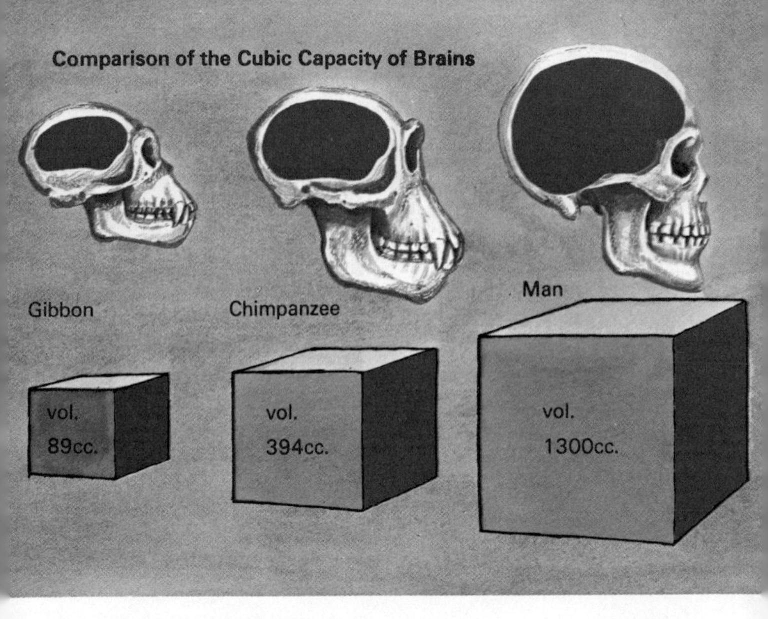

its remarkable capacity. Despite this, the modern human brain is not the largest brain known – those of the elephant and the whale are larger; but these are of course very large animals. It is therefore not only absolute size that is important, but also relative size in comparison with the body bulk of the animal.

If the brain sizes of a series of primates are compared a progressive increase in size and complexity can be seen, from the smaller prosimians to the larger apes and man. This can be paralleled by a similar increase in the complexity of their behavioural patterns.

During the known evolutionary history of man, an overall increase in brain size from 600–700 cc. to 1200–1300 cc. can be discerned. It is easy to speculate that this increase in overall brain size must correlate with greater intelligence and capacity for thought, but, in modern human populations, normally intelligent people may have brains ranging in size from 1000–2000 cc., and with this range there is no convincing evidence that intelligence is related to volume. Clearly other factors are involved, such as numbers of brain nerve cells and the complexity of their connections with one another.

THE PRE-HUMAN PHASE

The phases of human development

Direct evidence of the evolution of man from pre-human ancestors rests on the remains that exist in the fossil record. In the simplest terms four principal structural phases in this evolutionary process can be recognized; the pre-human phase, the australopithecines, the early human phase, the habilines, the late human phase, the pithecanthropines, and the modern human phase, which are the sapients. Controversy exists, however, over the validity of separating the first two phases. All of these men and ape-men lived during the Pleistocene Period and at the dawn of the Pleistocene Period, about two million years ago, a group from south and east Africa that were known as the australopithecines (the southern apes) were in existence.

As far as is known these hominids consisted of at least two populations; the one of small, light-bodied, plains-dwelling creatures, and the other of larger and more powerful creatures, who may have lived in a wetter, forested environment.

East African hominid fossil sites

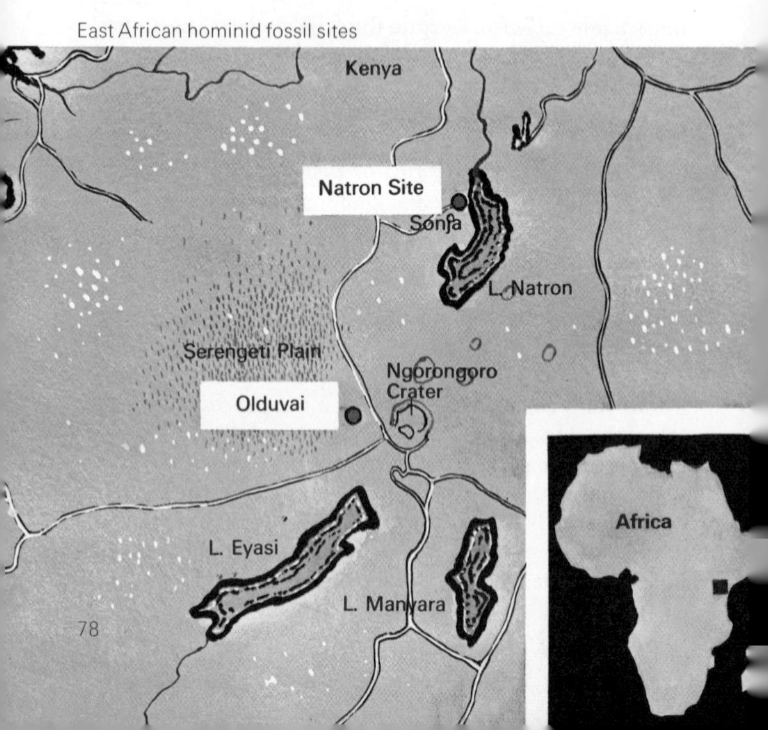

Although they were unknown before 1925, undoubted remains of australopithecines have been recovered from sites in the Republic of South Africa, Botswana and Tanzania. The southern sites consist of hard cave deposits discovered when the limestone was being quarried for cement manufacture but unfortunately most of the caves have collapsed so that no trace of occupation layers can be discerned. The bones are jumbled up with the remains of other members of the associated fauna so that the task of identification is made doubly difficult. Not only have the cave roofs fallen in but also solution cavities have formed under the original floors so that slumping of the roof and floor may further confuse the stratigraphy. Some primitive stone tools have been reported from one or two of the southern sites, but they are few in number and on this account they are not generally regarded as forming a true culture. However, claims have been made for an osteo-donto-keratic culture (bone, tooth and horn tools) in association with australopithecines in South Africa, although opinion is divided as to whether they are artifacts or simply carnivore-chewed remains.

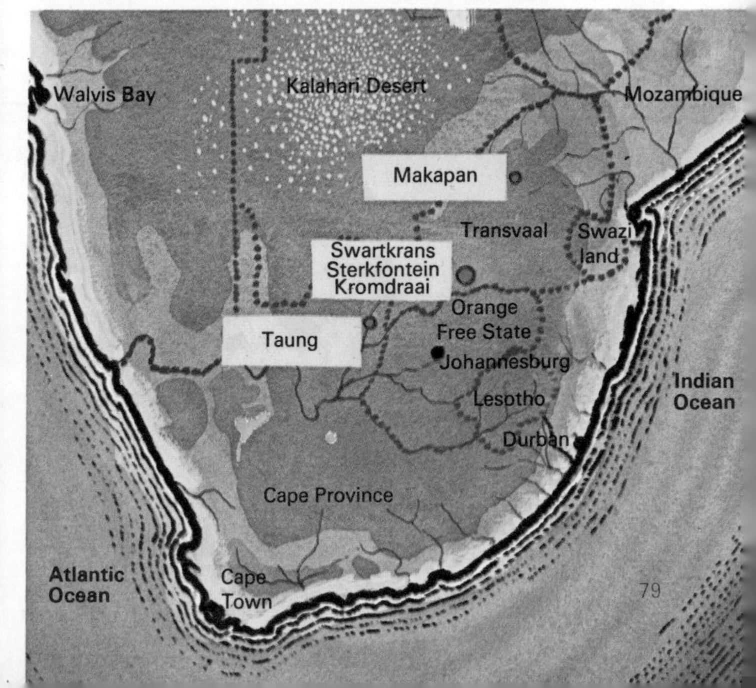

Australopithecines: a southern African map of distribution

The Taung site, Cape Province

In 1925, a quarryman who was blasting for limestone came across a small ape-like skull in a patch of sandy breccia which formed the infilling of an ancient cave in the face of an escarpment. This escarpment was part of the oldest of four limestone masses which form the valley of the Harts river. The specimen was drawn to the attention of Raymond Dart, an anatomist at the University of Witwatersrand. Dart immediately realized the importance of the find because he felt that, despite the superficially ape-like appearance of the skull, it had several hominid (human-like) features. No stone tools were found with the skull but numerous fossil mammalian bones were recognized; baboons, rodents, hyraces, and a bat were all identified, but no elephants or carnivores were found. Dating of the deposit has proved difficult but most authorities believe that the Taung skull is one of the earlier of the South African australopithecine specimens that is aged about one to one and a half million years old.

The specimen comprises the greater part of a juvenile skull, a broken jaw and a fossilized cast of the brain. The face of the skull and much of the base is intact, but a lot of the vault is missing. The jaw is broken but all the teeth are present, at

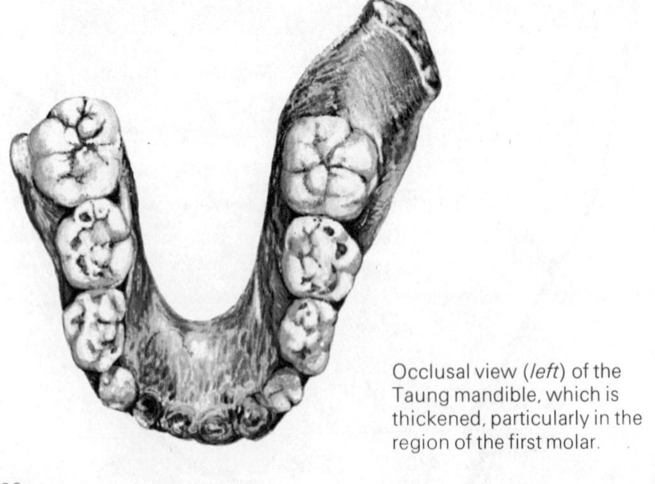

Occlusal view (*left*) of the Taung mandible, which is thickened, particularly in the region of the first molar.

least in part. The features which attracted Dart's attention are the rounded appearance of the cranium, the prominence of the forehead, the forward position of the foramen magnum, the regular, rounded outline of the dental arcade and the detailed anatomy of the molar teeth. The cranium is large for that of a juvenile ape, but small for a juvenile man. The foramen magnum, the large hole in the base of a skull which transmits the spinal cord, is set forward in the human position and thus indicates an upright stance. The rounded tooth row is unlike that of any ape and finally, the first permanent teeth bear an occlusal cusp arrangement, which is clearly hominid.

When Dart announced his discovery, there was much controversy, his opponents claiming that the Taung skull was only that of a fossil ape. It was not for many years after the discovery that this claim for the hominid status of *Australopithecus africanus* was fully vindicated by further finds at other sites in southern Africa.

The Taung skull, right lateral view (*below*)

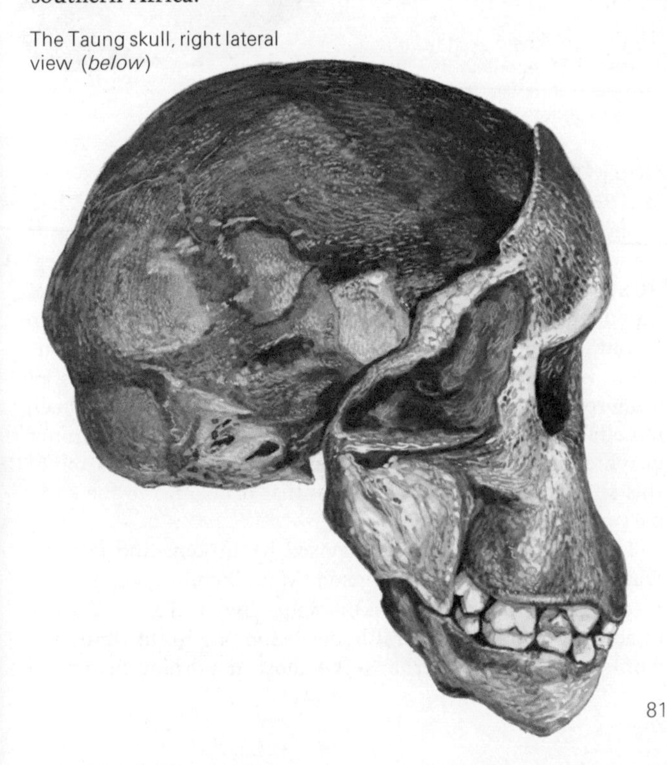

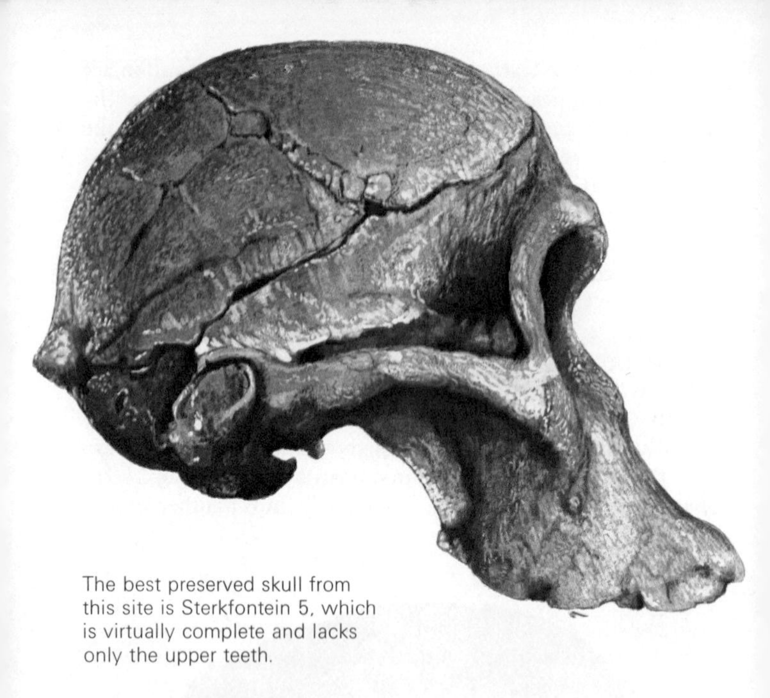

The best preserved skull from this site is Sterkfontein 5, which is virtually complete and lacks only the upper teeth.

Sterkfontein

In 1936, Dr. Robert Broome, a palaeontologist, who had supported Professor Dart in his battle over the hominid status of the Taung skull, found a new site not far from Johannesburg. At Sterkfontein there are a number of caves that have penetrated the hills, which are a formation of Pre-Cambrian dolomitic limestone. Over the years the caves have become filled with roof falls and have entombed the remains of numerous animals. This calcareous bone breccia has been mined for many years for burning in kilns for lime. The manner in which the deposit has been formed has made the dating of this site imprecise, but it is likely that it belongs to the early part of the sequence.

The excavations were supervised by Broome and later he was joined by Dr. John Robinson. Many hominid bones were recovered including a fine skull, some jaws and a lot of teeth. In addition to these, several bones belonging to the limbs and trunk were found, perhaps the most important being the greater part of a pelvis. A number of stone tools have been recovered from the excavation sites as well as bones of a mammalian fauna, which included insectivores, primates, numerous rodents and some carnivores.

The best skull (Sterkfontein 5) has many interesting features. The vault is rounded and has a moderate brow ridge, the foramen magnum is set forward and the ridge on the back of the skull is low. The teeth are all missing but the ridge of bone into which they were set is rounded and not rectangular as in the modern apes. In general the front teeth tend to be small, and the molar teeth large and irregular in shape. The upper limb bones are small, but not unlike those of modern man, while the pelvis is of particular interest since its features are a mixture of those found in man and the great apes. The blades of the hip bones are broad, flat and manlike in some respects, but the ischia (the projections of bone upon which we sit) are longer than those of man. The remaining limb bones include some fragments of the thigh bone which, although manlike, are much smaller.

In summary, the Sterkfontein australopithecines are small-brained ape-men whose limb bones suggest that they were upright and bipedal. They are grouped with the Taung skull as being representatives of *Australopithecus africanus*.

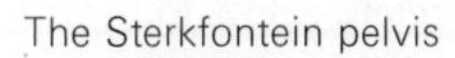
The Sterkfontein pelvis

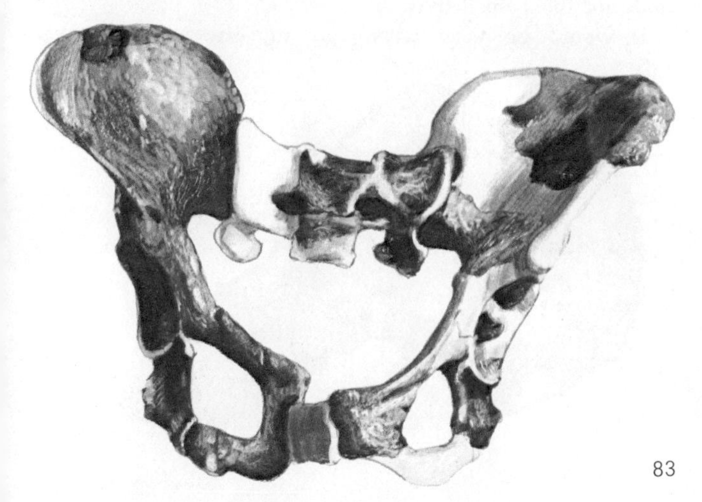

Makapansgat

Further remains of *Australopithecus africanus* have been found by Professor Raymond Dart at Makapansgat Limeworks Dump, Central Transvaal. These specimens consist of parts of skulls, a number of jaws, numerous teeth and several post-cranial bones. The dating of this site is probably similar to that of both Taung and Sterkfontein.

One of the best crania from Makapansgat was found in a block of stone which had been split. Fortunately the other half was found and the two parts were reunited. This skull is remarkably like that from Sterkfontein, although most of the face is missing.

The mandibles from this site comprise a range of examples from infancy through adulthood to senility. One of the best specimens is half an adult mandible which contains the canine premolar and molar teeth. In general, the jaws are short with rounded dental arcades, divergent tooth rows and no chins. The teeth are similar to those that were recovered from Sterkfontein.

The Makapansgat limb bones consist of two juvenile ilia (hip blades), an ischium, part of a humerus (upper arm bone) and a few other fragments. The ilia are similar in many respects to those from Sterkfontein. The blade of each is twisted into an S-shape when seen from above and the breadth of the blade recalls the human shape. The ilium of a modern ape is long and does not show the S-twist.

It would be very wrong to suggest, however, that

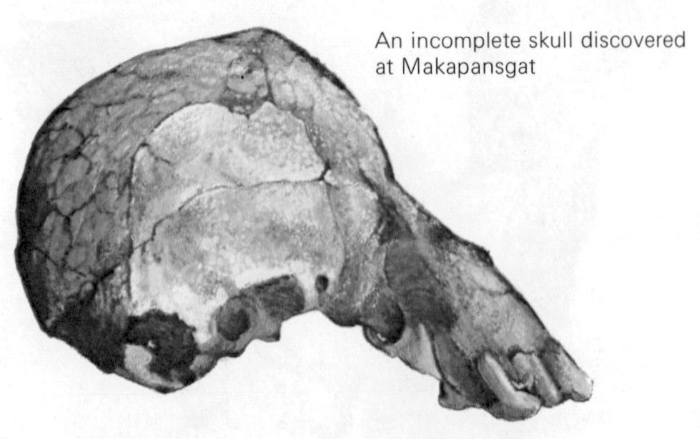

An incomplete skull discovered at Makapansgat

the smaller australopithecine pelvis is the same as that of modern man for there are several differences. The anterior spine of the australopithecine hip bone reaches well forward whereas that of modern man does not. Similarly the strut-like iliac pillar of man is feebly represented in the ape-men. These and other differences have led to controversy over the possible forms of locomotion of these creatures. One school of thought holds that they were bipedal, upright walkers, but not as good as modern man. The other school, much the smaller, maintains that the evidence is insufficient to tell what their gait was like.

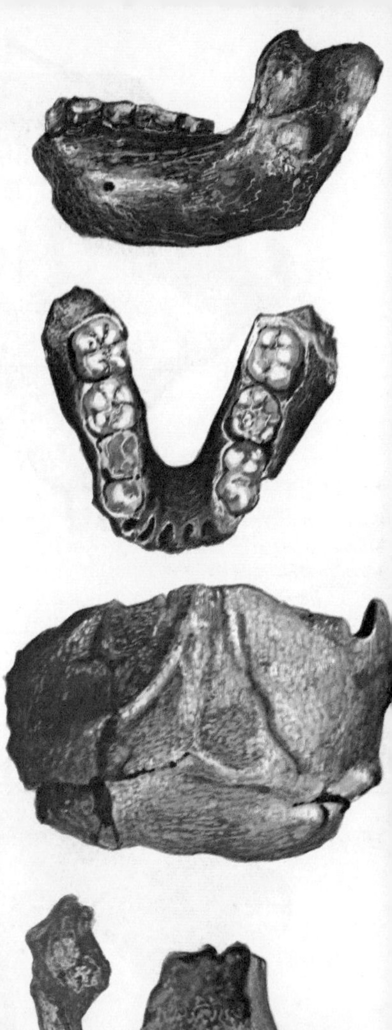

Mandibles (*top*), a cranial fragment (*right*) and two juvenile ilia and an ischial fragment (*below*) from Makapansgat

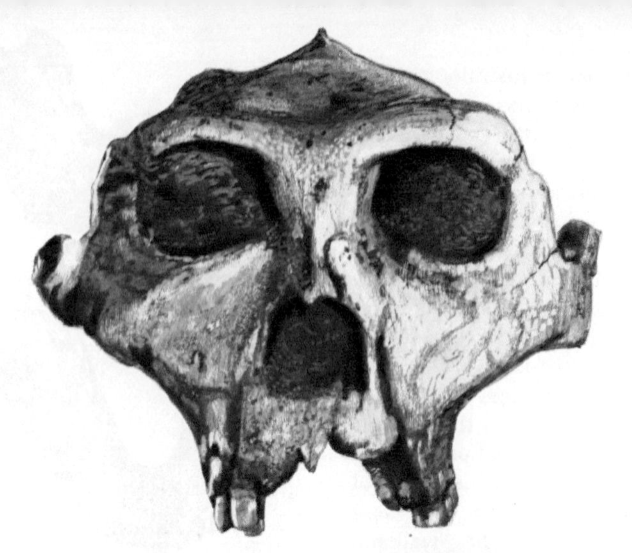

The best skull from Swartkrans is marked by stout ridges for muscle attachments. The brain case is small while the face is large.

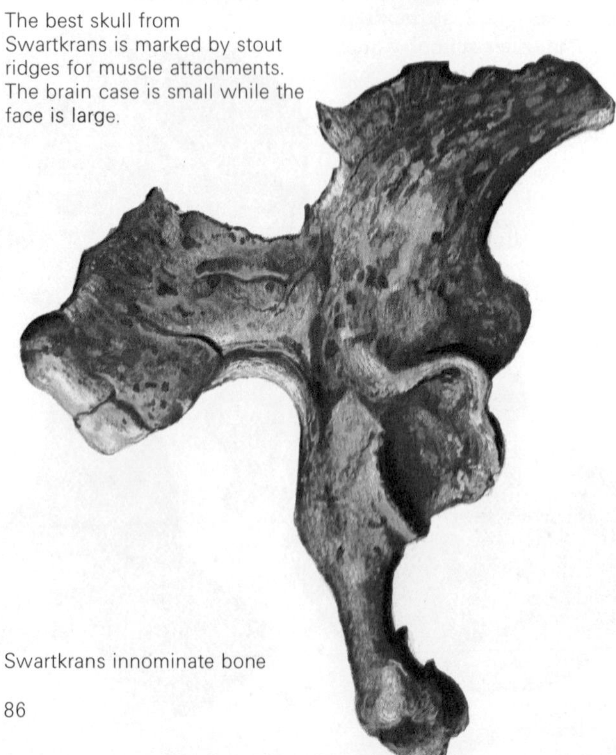

Swartkrans innominate bone

Swartkrans

While working in the Transvaal, Broome and Robinson conducted extensive excavations at a site near to Sterkfontein. This site is all that is left of a large cavern in the limestone; the outer part of its roof has been removed by erosion but the inner cave is still covered by a layer of rock. The cave filling is made of pink breccia underlying younger, brown breccia and it was the older pink breccia that contained hominid fossils. It seems likely that the pink breccia is younger than the deposits at Taung, Sterkfontein or Makapansgat.

The hominids obtained from this site, and the next site (Kromdraai), are very different from the australopithecines previously described. These later creatures are much larger, powerful animals, with rugged bones, big teeth and heavy jaws.

The best Swartkrans cranium is a little distorted, but includes a well-marked crest on the top, the sagittal crest, for the big jaw muscles. The bony palate is large and a lower jaw from the same site is very tall and robust.

The jaw has no chin and is buttressed by a torus or ridge inside the teeth. Over 250 teeth are known from Swartkrans, illustrating a number of features which distinguish them. Few limb bones have been found at Swartkrans but part of a pelvis has been discovered. This hip bone has some of the features of the smaller Sterkfontein pelvis and seems intermediate in form between that of the modern apes and man.

One of the many mandibles discovered at Swartkrans (*below*). The front teeth are disproportionately small in comparison with large molars.

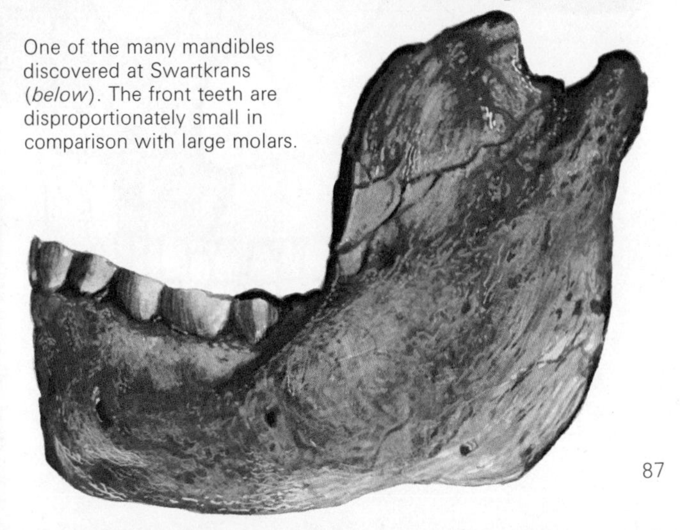

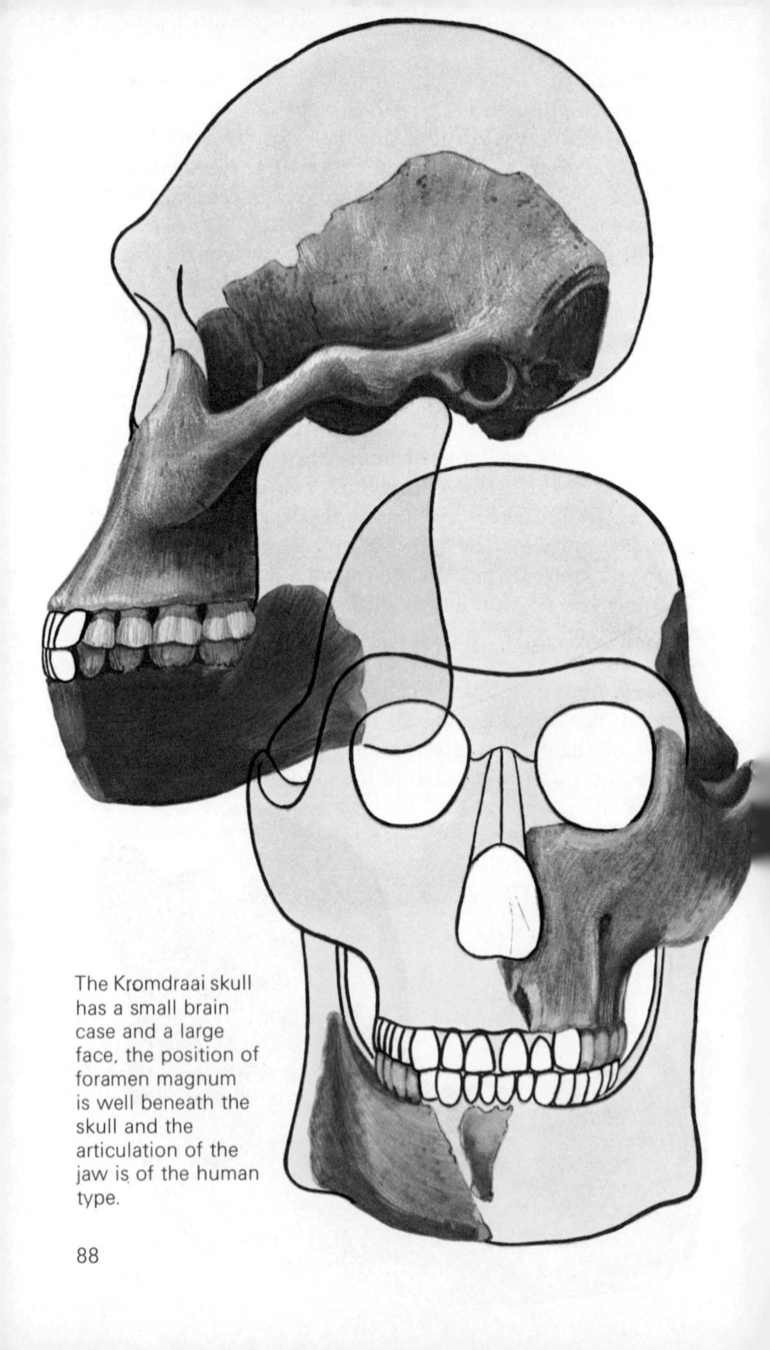

The Kromdraai skull has a small brain case and a large face, the position of foramen magnum is well beneath the skull and the articulation of the jaw is of the human type.

Kromdraai: Close to Swartkrans and Sterkfontein

A schoolboy, Gert Terblanche, found some fossils which were later shown to Robert Broome. He recognized their importance and was taken to the site by the boy. Once again this proved to be a cave site in Pre-Cambrian dolomitic limestone where the roof and walls of the cave had weathered away leaving the cave filling exposed on the surface. Dating is difficult in this area but it is believed that the Kromdraai site is the same age as Swartkrans. From Kromdraai B, the name of the discovery site, a skull was recovered. Part of a jaw shows it is stout and has large teeth and the characteristics of the teeth are not unlike those of the larger australopithecine from Swartkrans. An incomplete set of milk teeth were also found at Kromdraai and the first milk molars are remarkably human in their shape and cusp arrangement.

The limb bones from Kromdraai are fragmentary but important. There is the lower end of a humerus which is very human in its general shape, a small piece of the ulna, a forearm bone, and a talus, one of the small ankle bones.

Kromdraai Limb Bones

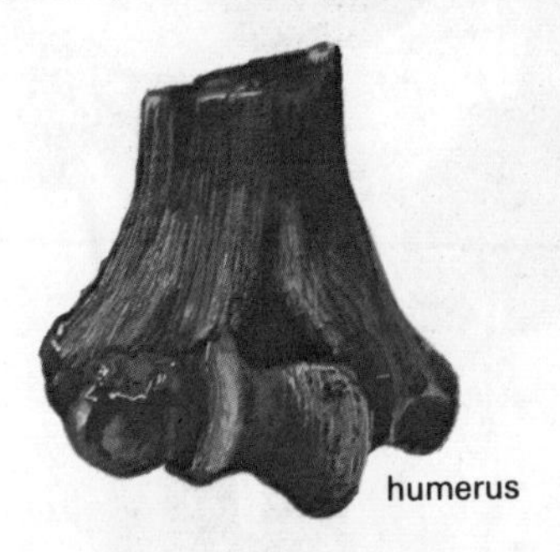

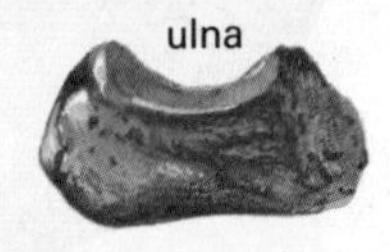

Olduvai Gorge, Tanzania

The best australopithecine site outside South Africa is to be found in East Africa at Olduvai Gorge (Bed I). Here, in deposits dated at 1,750,000 years BP, Mrs Mary Leakey found a cranium belonging to a large australopithecine '*Zinjanthropus*'. The skull was found embedded in a living floor sited near the margin of an ancient lake. Associated with the specimen were a large number of broken mammalian bones, and the remains of small amphibia and fish.

The most striking features of the skull are the size of the face

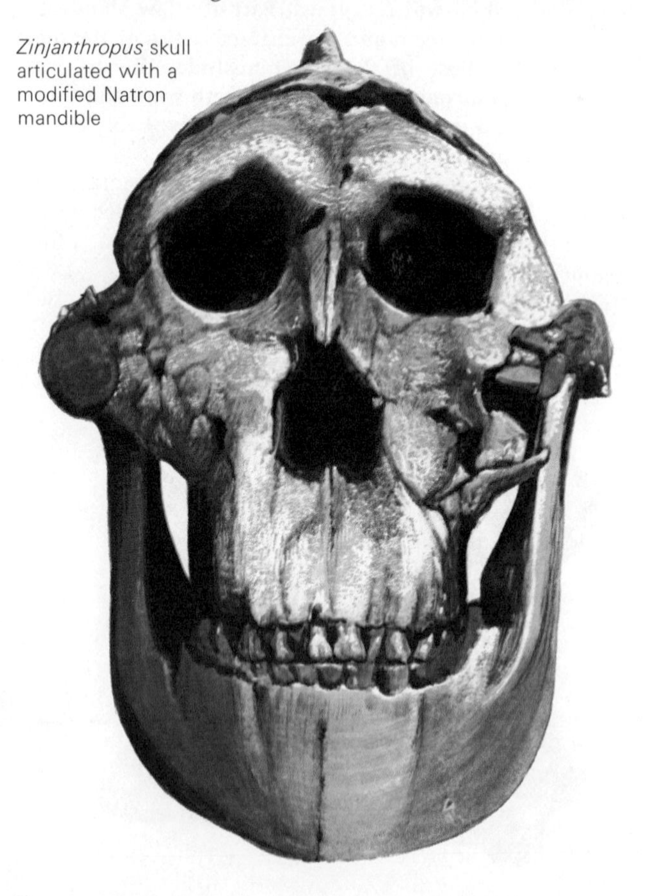

Zinjanthropus skull articulated with a modified Natron mandible

in comparison with the volume of the brain case and the marked development of ridges round the cranium, such as the sagittal, occipital and supra-mastoid crests. The extent of these muscle attachments suggests a very large and heavy jaw. The teeth of this specimen are enormous and, although worn, resemble the other australopithecine types. Subsequently this specimen has been named *Australopithecus boisei*.

Lake Natron

By a curious chance an almost perfect, large australopithecine mandible was found in 1964 by Richard Leakey at Peninj, to the west of Lake Natron in Tanzania. It contains the complete adult lower dentition and the resemblance of this Natron jaw to the best jaw from Swartkrans is remarkable. Both examples have high rami, robust bodies and large teeth. The typical disproportion of front and back teeth, and the squared-off incisor row argue strongly that this bone has unmistakably australopithecine affinities.

The Natron jaw

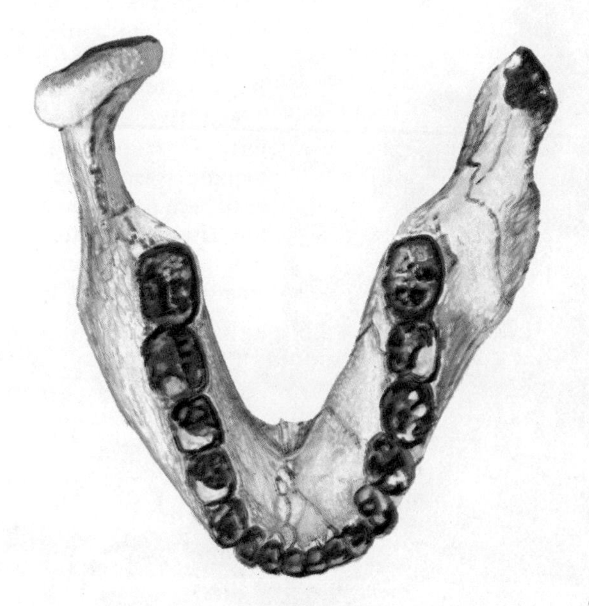

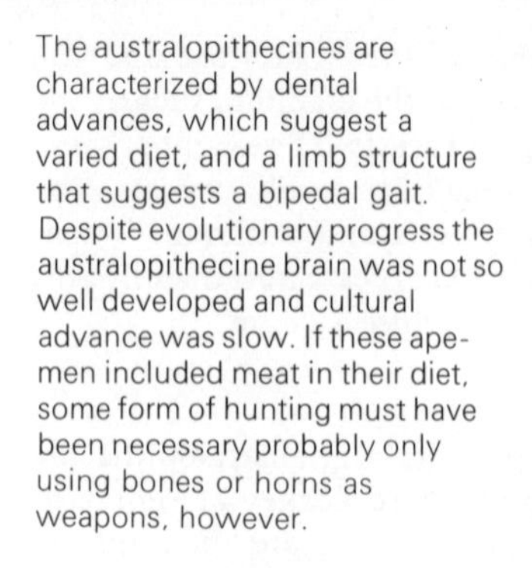

The australopithecines are characterized by dental advances, which suggest a varied diet, and a limb structure that suggests a bipedal gait. Despite evolutionary progress the australopithecine brain was not so well developed and cultural advance was slow. If these ape-men included meat in their diet, some form of hunting must have been necessary probably only using bones or horns as weapons, however.

The australopithecines

The australopithecines can thus be divided into two principal groups. One group small in terms of body size, the other large. The arguments of the past, as to their right to hominid status, seems to be settled.

It is widely accepted that the remains from Taung, Sterkfontein and Makapansgat all represent one species, *Australopithecus africanus*. These animals were small-brained, lightly built, agile and probably bipedal ape-men who lived in and around the limestone caves of the Transvaal about a million years ago. Their teeth indicate that they ate a diet which might have included meat. If so, it is likely that they would have engaged in some form of hunting, and that tool or weapon use would have been part of their way of life. The evidence for cultural tool-making is scanty apart from the claims of Professor

Dart for an elaborate bone-tooth-horn culture.

The larger australopithecines from Swartkrans, Kromdraai and Olduvai are also small-brained, exceptionally large-toothed animals who lived in a more forested environment. The size and condition of the teeth has suggested that these ape-men were predominantly vegetarian, eating roots, bulbs and stems although meat eating is by no means precluded by their dentition. Their classification is still unsettled but a few authorities maintain that they belong to a separate genus, *Paranthropus*, while others would include them in the genus *Australopithecus*.

Whatever the outcome of this particular argument, the australopithecines of Africa represent a 'pre-human phase' of hominid evolution. Despite this progress, the australopithecine brain has lagged behind and shows little tendency towards expansion.

Features such as the shape of the cranium, the position of the foramen magnum, the anatomy of the teeth and the post-cranial evidence of posture and gait, have provided convincing evidence of the hominid status of the australopithecines.

THE EARLY HUMAN PHASE

Olduvai Gorge

The fossils that provide the evidence upon which this structural stage depends also come from Africa, from Olduvai Gorge in Tanzania. This site is situated on the Serengeti Plain, about thirty miles north of Lake Eyasi. The gorge is an offshoot of the Great Rift Valley which splits east Africa from north to south; it is about 300 feet deep and consists of well stratified deposits covering approximately the last two million years. The fossil bearing layers rest on a lava base and have been numbered Beds I-V from below upwards. The rocks are volcanic in origin and the strata were laid down in the beds of lakes, streams, or by the action of the wind. The conditions for fossilization were excellent.

About two feet below the layer in which the large australopithecine skull was found, but 300 yards away, new material was recovered by Dr L. S. B. Leakey and his team in 1960. The remains consist of two groups of hominid fossil bones belong-

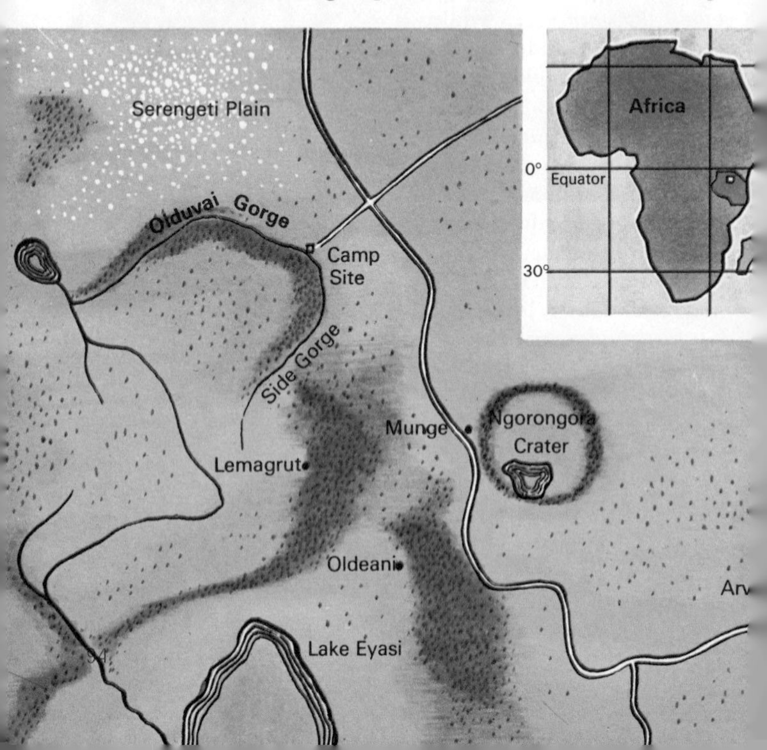

ing to at least two individuals, one adult and the other juvenile.

The Bed I early human material consists of two juvenile parietal bones (part of the skull vault), a jaw, a clavicle (collar bone), parts of two hands and an almost complete left foot. Examination of this material has led Dr Leakey and his colleagues to attribute these bones to a new species of man, *Homo habilis* (able, or skilful man.) Thus *Homo habilis* is the earliest known hominid to be accredited with the ability to manufacture simple tools or weapons to a consistent pattern. The layer in which the bones and tools were found has been dated by more than one method at about 1,750,000 years BP.

The skull bones are thin and incomplete but restoration of the biparietal arch was possible and from this restoration an estimate of the capacity of the complete skull has been made. It is higher than those of the known australopithecines although its growth is incomplete. The jaw is also immature and somewhat crushed but it is stout and contains most of the permanent teeth. The incisors are hominid and the pre-molars are said to be elongated from front to back.

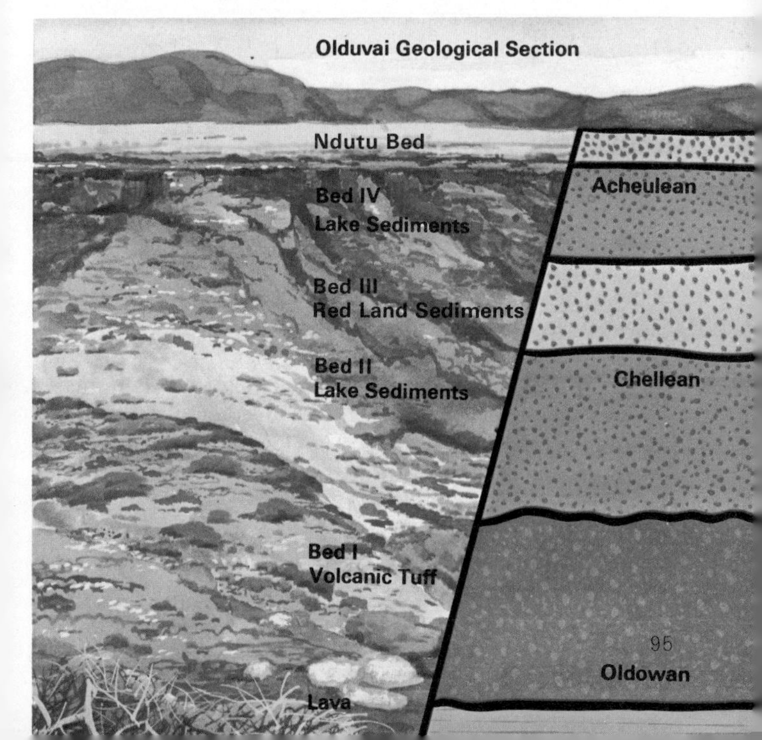

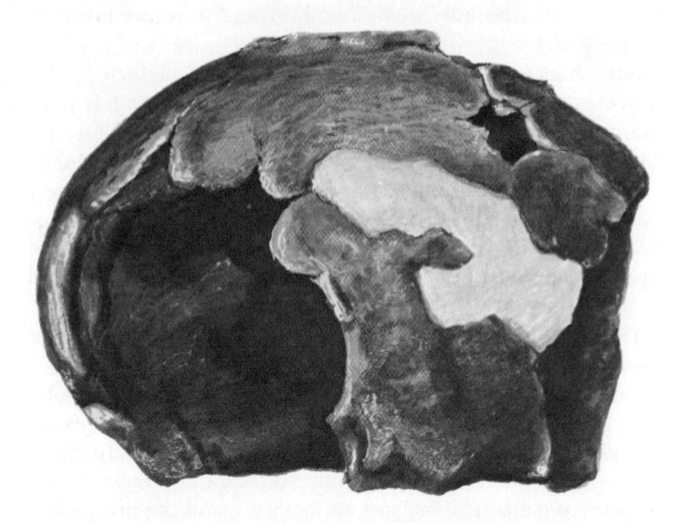

The skull from Olduvai, Bed II

Homo habilis

At the same Olduvai site there are also about fifteen hand bones from two individuals; one adult and one immature. The finger bones are small and curved except for the thumb which is of particular interest since it is flattened at the tip. This is important because it shows that the thumb could have helped to make a human type of grip.

The hand bones from the Bed I site cannot be very closely matched with those of any known hominid species. They do, however, bear a closer resemblance to a present-day juvenile gorilla and also to modern man, rather than to any modern ape.

The foot is almost complete and comprises twelve bones. When it was articulated, that is, put together as in life, it was found to be damaged by a predator or scavenger. The toes and the projection of the heel are missing, probably gnawed away, since the talus bears quite distinct tooth marks on its upper surface. This foot is an exciting find, since its shape can tell us a lot about the way its owner stood and walked. For example, the position of the big toe is indicated by its metatarsal, and in this case it is beside the other toes as in a modern human walking foot; in addition the distribution of strength in the metatarsals conforms to the human pattern, and the formation of arches is human in type.

As well as the foot, two leg bones and a toe bone have been obtained from Bed I. The leg bones, a tibia and fibula, show by their lower ends that they probably belonged to a biped; from their size it seems not unlikely that they should belong to *Homo habilis* rather than to *'Zinjanthropus'*, with which they were found. The toe bone is short, flattened and tip-tilted as are those of modern bipedal walkers.

In summary, the evidence of the leg and foot bones from Bed I, Olduvai Gorge, indicates that men of the early human phase were capable of standing erect and walking on two legs, but not as efficiently as present-day man.

Since the discovery of this material, other finds have been made at Olduvai supporting these theories. Material found most recently at Olduvai comes from Lower Bed II, a slightly more recent layer than Bed I. The vault of a small skull and parts of both upper and lower jaws were found, also a second skull vault and some teeth. These bones are not yet fully described but it has been suggested that they have some advanced features.

The skull from Bed II at Olduvai Gorge, compared with the skull of modern man

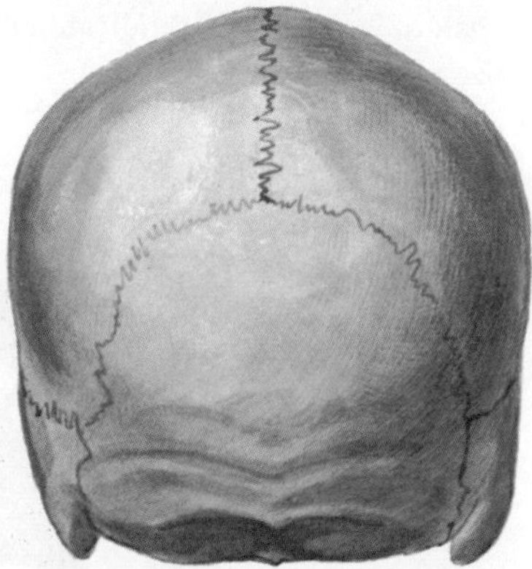

The new species

The announcement of the discovery of a new species of extinct man by Dr Leakey and his co-workers aroused a controversy which is not yet settled. By creating a new species within the genus *Homo* it was asserted that the habiline material cannot be matched by any other known hominid bones and that it fell outside the variation ranges of other species. In effect this means that the habilines were thought to be so much in advance of the australopithecines that they could not be classified with them. The principal grounds put forward for the distinctiveness of *Homo habilis* are on the basis of cranial and dental features, but some evidence from the structure of the limbs and the cultural activity of these hominids has also been used.

The skull volume of the *Homo habilis* type skull has been calculated from the reconstruction of the vault and estimated at about 670 cc. This is larger than that of any known australopithecine, but smaller than that of any example of the next group, the pithecanthropines.

The teeth and jaws are said to be distinctive in a number of ways, the curve of the habiline jaw is more rounded than the australopithecine jaw and the measurements of the teeth are said to be distinct in the two groups.

The limb bones of *Homo habilis* are unquestionably advanced in their anatomy in that there is evidence of a hand capable of making at least a power grip and a foot, a toe and probably a pair of leg bones that indicate a fairly advanced form of bipedal

Reconstruction of the Olduvai foot

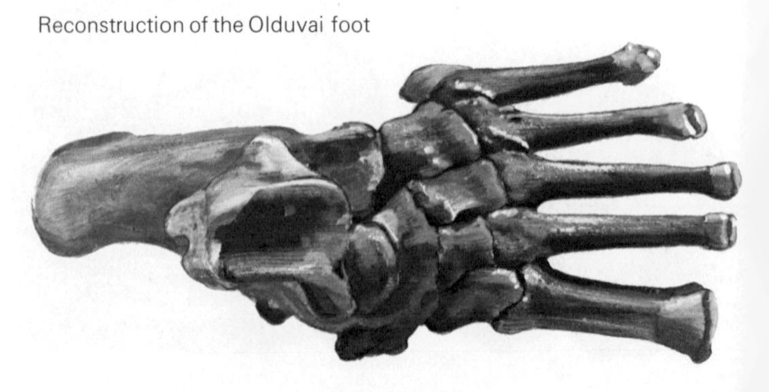

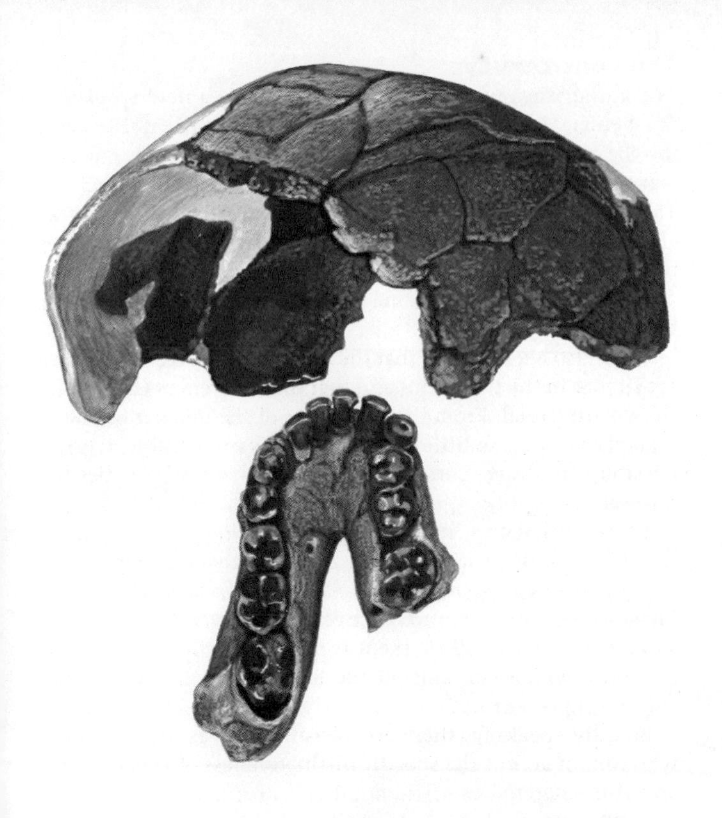

Homo habilis, skull vault (*top*) and jaw

walking in their owners. The only comparable material is one small ankle bone from Kromdraai, which may have belonged to the larger australopithecine, that shows some indication of having a divergent great toe and thus possibly a grasping foot. Two hand bones and a small wrist bone are all that is known of the australopithecine hand and it is very difficult to assess functional capability from these scanty remains.

The stone tools from Olduvai form a definite culture. The tools are simple but have been found in sufficient quantity, associated with *Homo habilis* remains, to permit the view that *Homo habilis* was a cultural tool-maker.

The controversy

The authorities, who oppose the creation of a new species of the genus *Homo* to include Olduvai bones, do not accept, for the most part, that sufficient evidence has been put forward for their distinctiveness from the australopithecines. They maintain that the brain size of *Homo habilis* will eventually prove to be within the range of variation of the australopithecines when this variation is known. To date, however, only seven australopithecine skulls have been estimated for brain size.

It has also been argued that the features of the teeth and jaws are similar in the two forms and that the differences that can be shown are trivial. From the viewpoint of the limbs it has been argued that since so little of the material is comparable, it is not possible to draw conclusions about their similarities or differences.

These authorities would suggest that the *Homo habilis* is simply a local variant of the australopithecine population known to have existed in sub-Saharan Africa in the Lower Pleistocene, possibly in advance of the Transvaal forms but not so advanced as to allow them to join the genus *Homo*; that grouping which contains all the known populations of true man living or extinct.

Broadly speaking, there are two alternatives open to those who cannot accept the specific distinctiveness of *Homo habilis* and *Australopithecus africanus*. These are:

(i) That *Homo habilis* should be combined with *Australopithecus africanus*.
That is to say that neither belongs to the genus *Homo*, both belong to the genus *Australopithecus* and the name *'habilis'* should be eliminated.

(ii) That *Australopithecus africanus* should be combined with *Homo habilis*.
That is to say that they both belong to the genus *Homo* and again the name *'habilis'* should be eliminated since the specific name *'africanus'* would have priority. Thus both would belong to *Homo africanus*.

Those who would maintain that these two forms are conspecific have to accept a definition of the genus *Homo* or of the genus *Australopithecus*, which can accommodate both creatures. Their argument is over generic features which are far less easy to identify than those which distinguish species.

In summary let it be clearly understood that most of the controversy is over the naming and classification of these important fossil finds. It is apparent that there were, in East Africa, a group of early hominids, who lived over $1\frac{1}{2}$ million years ago and were bipedal and were capable of making stone tools. In these respects at least, these are some of the earliest man-like creatures that we know. When the East African habiline and australopithecine material has been studied more closely it is probable that a lot more light will be thrown on the relationship of these early hominids.

Until such time as all the various finds are fully evaluated, the Transvaal australopithecines provide evidence of a 'pre-human' phase of evolution and the habilines, evidence of an 'early human' phase of evolution. Both groups have made great advances in the modification of the teeth and the post-cranial skeleton, but the brain has only developed slightly. Only more study and more material will clarify the issues involved in their proper understanding.

Some authorities suggest that *Homo habilis* is a variant of the australopithecine population and not advanced enough to join the genus *Homo*.

THE LATE HUMAN PHASE

The pithecanthropines

The pithecanthropines form the group of fossil men who lived during the Middle Pleistocene, about half a million years ago. It is fortunate that they were widely distributed in the Old World, specimens of these men (*Homo erectus*) having been recovered from Java, China, Africa and possibly Europe. There is no doubt about their claim to humanity since they fulfil all the anatomical and cultural criteria required of this status.

Map illustrating the location of hominid fossil sites in Java

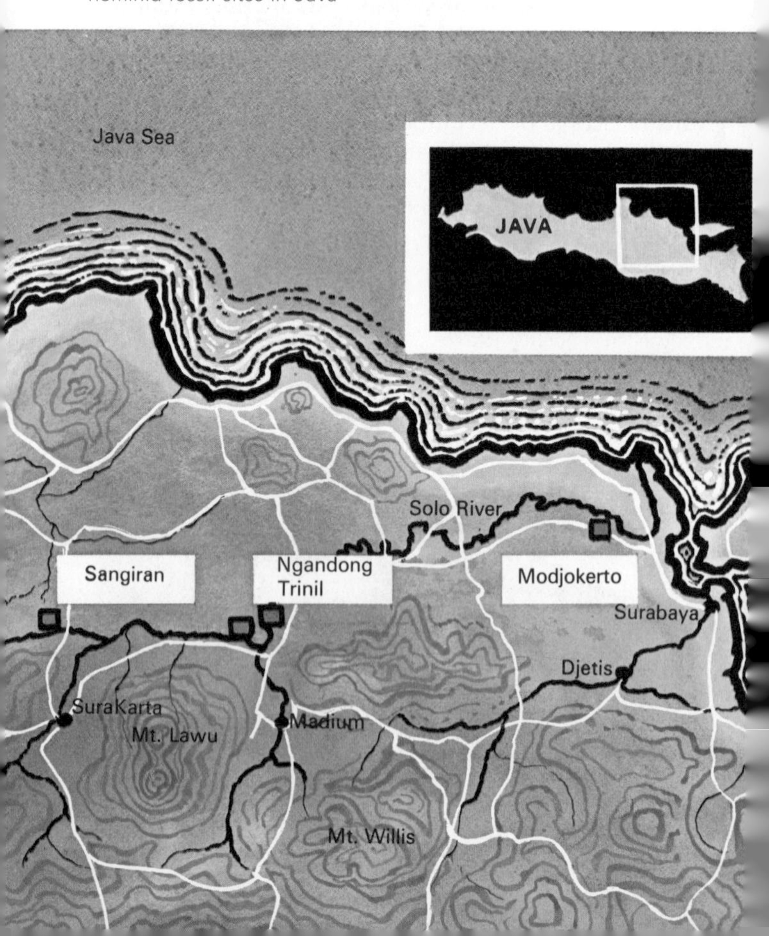

Lateral view of the Trinil skull cap and (*below*) an internal view

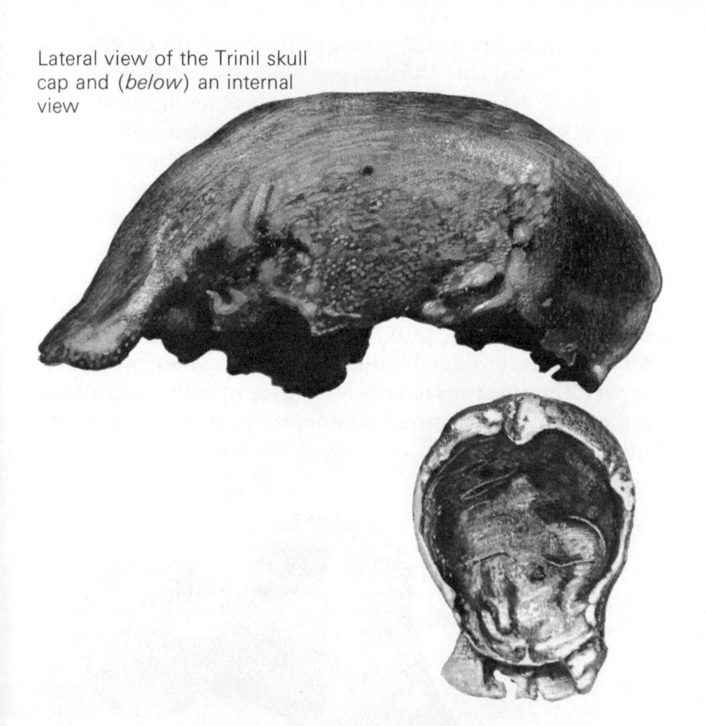

Java man *(Homo erectus erectus)*

This specimen was one of the earliest of fossil men to be found. Dubois, a Dutch anatomist, threw up his job and travelled to the Far East with the intention of looking for fossil human remains. It was to his credit that he succeeded in the year 1891. At a place called Trinil, in Central Java, on the banks of the river Solo, Dubois found a skull cap embedded in the Middle Pleistocene deposits. Not far away from this he also found a complete thigh bone.

These finds caused furore, because the thigh bone was modern in form while the skull cap appeared primitive. It was thick, heavy and flattened in front and the ridge above the eyes was prominent and ape-like. The combination of an ape-like skull and a modern femur led to it being given the name of *Pithecanthropus erectus* (the erect ape-man). This name was changed later to *Homo erectus* (erect man).

The femur of this Java example of *Homo erectus erectus* is of particular interest since it bears one of the earliest examples of the results of disease in man. The shaft is straight but near to the upper end of the bone there is an excrescence which has been perfectly fossilized. It is the end result of a condition that is known as *myositis ossificans*. Perhaps as the result of an injury to the leg, the processes of bone formation have spread into the surrounding tissues thus leaving clear evidence of a disease process in fossil man.

The modern appearance of the femur contrasts greatly with the archaic form of the skull, so much so in fact that for many years there was considerable doubt as to their contemporaneity. However, fluorine tests have been applied to both the skull cap and the femur and removed all doubts. They are undeniably of the same age.

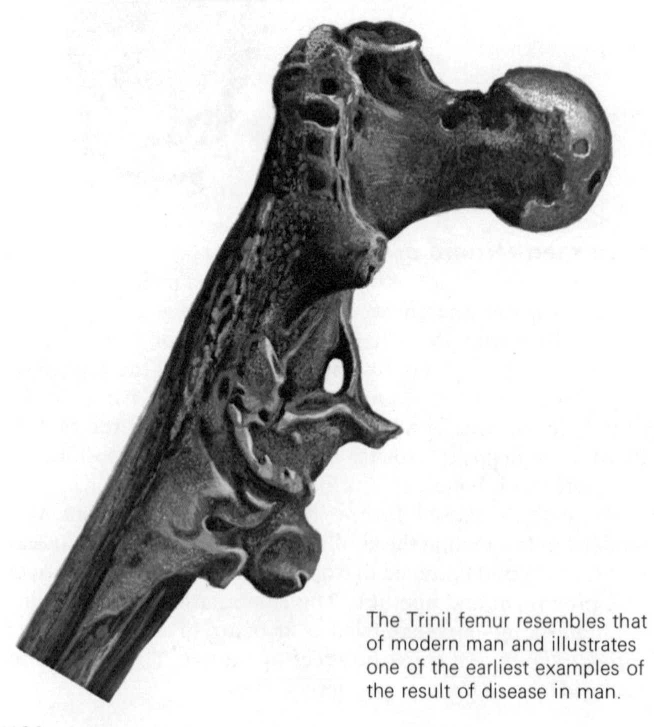

The Trinil femur resembles that of modern man and illustrates one of the earliest examples of the result of disease in man.

The Sangiran remains

Over forty years later at a place called Sangiran, not far from Trinil, Professor von Koenigswald uncovered another skull that was almost identical to the one found by Dubois at Trinil. The remains were found in the Djetis layer that underlies the Trinil deposits at Sangiran. Fossil mammalian fauna discovered in the same deposit as the skull included a primitive ox and a sabre-toothed cat.

This second skull is rather better preserved than the first and has some indication of the position of the foramen magnum, showing that the opening is set forward and suggesting that *Homo erectus* was indeed habitually upright. The lower portions of both maxillae are preserved but were crushed before fossilization and therefore are distorted. All the teeth are present and not very worn. Later finds at this site, but from a deeper layer, include an upper jaw and the back of another rather large skull.

The Sangiran maxilla has been well preserved but was crushed before fossilization and is thus distorted.

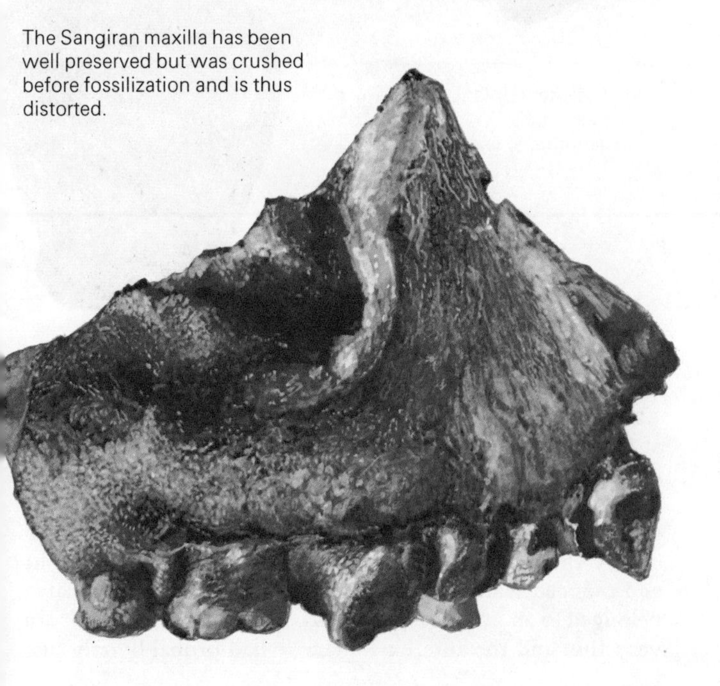

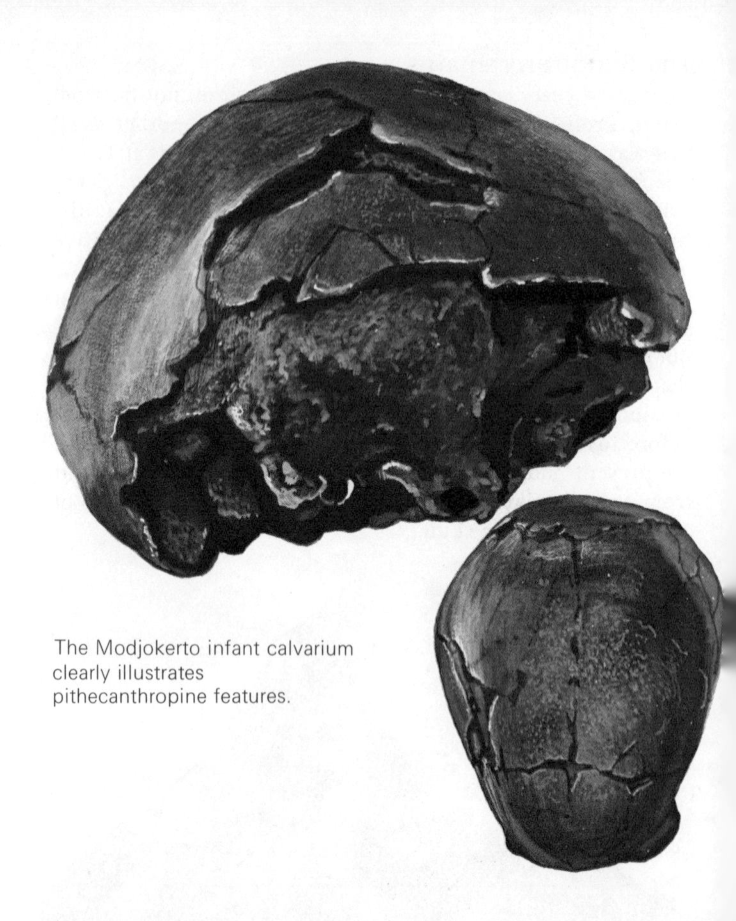

The Modjokerto infant calvarium clearly illustrates pithecanthropine features.

Other Java discoveries

At about the same time, 1936, an infant skull was discovered at Modjokerto, west Java, in a bed of Pleistocene river sands and marine sediments that are situated at a lower stratigraphic level than that containing the Trinil remains that have already been described. This little skull lacks its face but its outline clearly shows pithecanthropine features. The flattened front and reduced width behind the orbits strongly suggest that it belonged to an infant *Homo erectus*. The bones of the vault are very thin and the anterior fontanelle had probably only just

The second *Meganthropus* mandibular fragment has three large teeth and the size of the jaw is larger than any example from modern man.

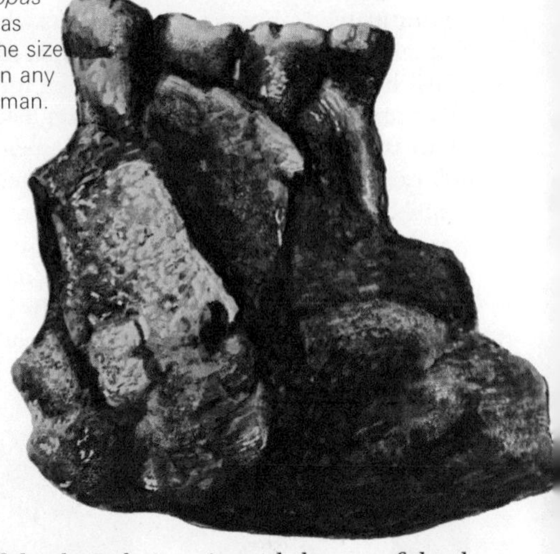

closed at the time of death. Dubois estimated the age of death as about two years.

Recently more *Homo erectus* material has been recovered from the fossil sites in Java and these include the back of a fine skull and further jaw fragments.

Just before the outbreak of the 1939–1945 war, two mandibular fragments were found by von Koenigswald and his assistants again at Sangiran. These fragments were massive and bore several enormous teeth. There is no doubt that they are remains of hominid jaws but they are considerably larger than any example of modern man and are only equalled by a few modern gorillas and there is, therefore, disagreement over the group to which they belong. Several authorities believe that they should belong to an early form of *Homo erectus* but Professor J. T. Robinson, an expert on the australopithecines, firmly believes these remains belong to a far eastern representative of the larger australopithecine of South Africa. A third and almost identical fragment of jaw was found nearby in 1945.

Whatever the final outcome is, it is clear that there were outsize hominids living in the far east at about the same time as *Homo erectus*. They have been named *Meganthropus palaeojavanicus* (giant old-Java man).

Peking man

Peking man was discovered in deposits which form the infilling of a large cavern in the limestone hills near Peking in China. Fossil bones and stone tools had been found for some time at this site but it was not until Davidson Black examined a single tooth that he was convinced that it belonged to fossil man. Boldly, on the evidence of this tooth alone, he created the genus *Sinanthropus* (China man). The cliff face at the principal site was divided into sections by Black and in the cave filling most of the discoveries were made. Much more material came to light at a later date including skulls, jaws, teeth and some limb bones. The deposits are of Middle Pleistocene Age and contain

The site of Black's excavations near Peking

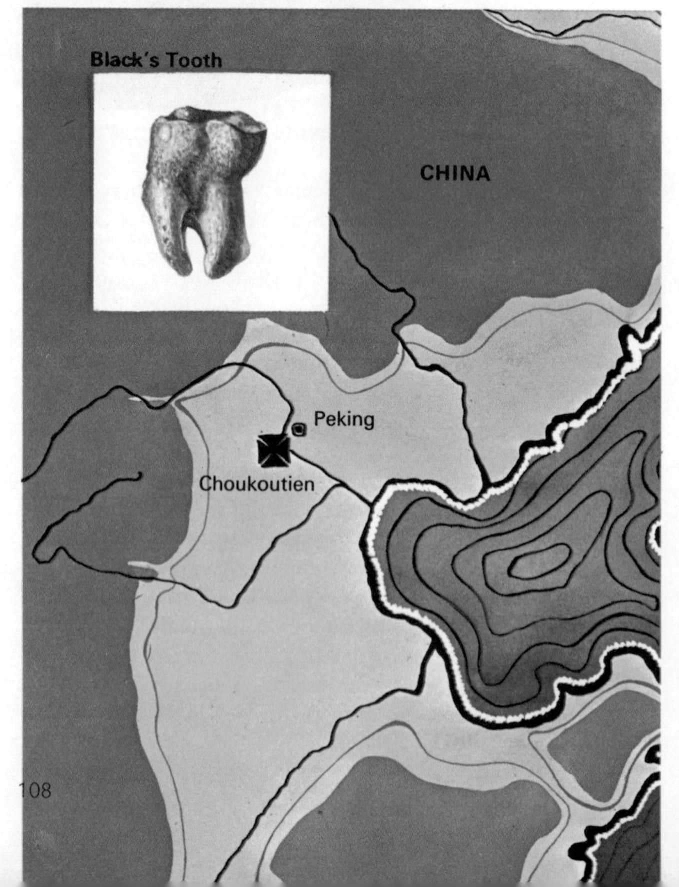

the remains of many large mammals, presumably brought back to the cave for food.

Fourteen incomplete skulls were recovered whose features recall the skulls from Java. The brow ridge is prominent and the braincase flattened, and seen from the front the widest part of the skull is set low on the side of the head. This feature is characteristic of *Homo erectus*, the species to which these remains are attributed. The jaws have several interesting features such as their rounded tooth row and the complete absence of a bony chin. The brain size of this group varies from about 900 cc.–1200 cc. The limb bones are broken and very small and they are similar to those of modern man.

The cliff face of Dragon Hill near Choukoutien

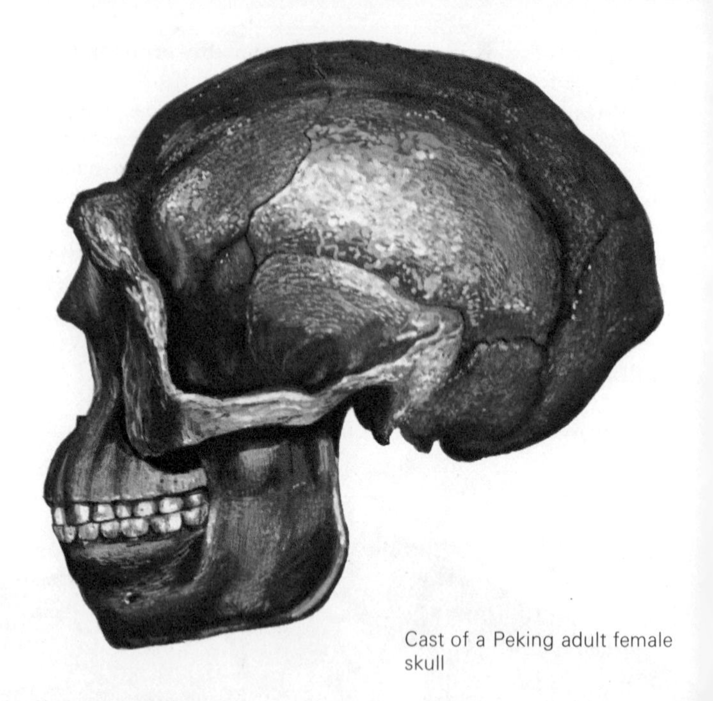

Cast of a Peking adult female skull

Examination of these remains by means of measurements gave a new insight into the structure of the population of the Middle Pleistocene hominids since it was shown that the adult teeth were of two principal sizes. This fact has been taken as evidence that, like many other primates, *Homo erectus* males were probably larger than females. The Peking teeth show several other interesting characteristics. In general the teeth are robust and particularly wrinkled. The incisors tended to be 'shovel-shaped', while the canines, premolars and molars almost all have a collar of enamel around the crown of the tooth known as a cingulum. Frequently the pulp cavity of the molars and premolars is enlarged, a condition known as taurodontism (bull-tooth). In Peking man the second permanent molar erupted before the premolars and the canines, unlike the order of modern man.

Finally the Peking site is of particular interest since the cave deposits contained clear evidence of the use of fire. Ash heaps were traced down through the layers for as much as eighteen feet showing that Peking man occupied the cave for very long periods and that the hearth was carefully tended. A fire at the mouth of a cave serves several purposes for not only does it warm, light and cook, but also it will ward off wild animals. The harnessing of fire was a major cultural breakthrough for mankind. In addition the stone chopper tools found at the site were sufficient in number to form a culture in advance of the Oldowan tools known from Bed I at Olduvai.

Peking man is known to have used fire.

North Africa

Further evidence of the Middle Pleistocene pithecanthropine group has been found at a series of North African sites, most of which are near the coastline. The principal site is at Ternifine, near Mascara in Algeria. Excavation of the bed of an ancient lake produced a varied group of mammalian fossils including hippopotamus, elephant, camel, and a giant baboon. There were also large quantities of stone tools, primitive hand-axes and scrapers. The hominid bones that were found at Ternifine all belong to the skull and comprise three jaws and a small parietal bone.

The Ternifine jaws show remarkable similarities to those known from Peking and the lack of chin, shape of the tooth row as well as dental resemblances, all suggest a close anatomical relationship.

Hominid Fossil Sites in North-West Africa

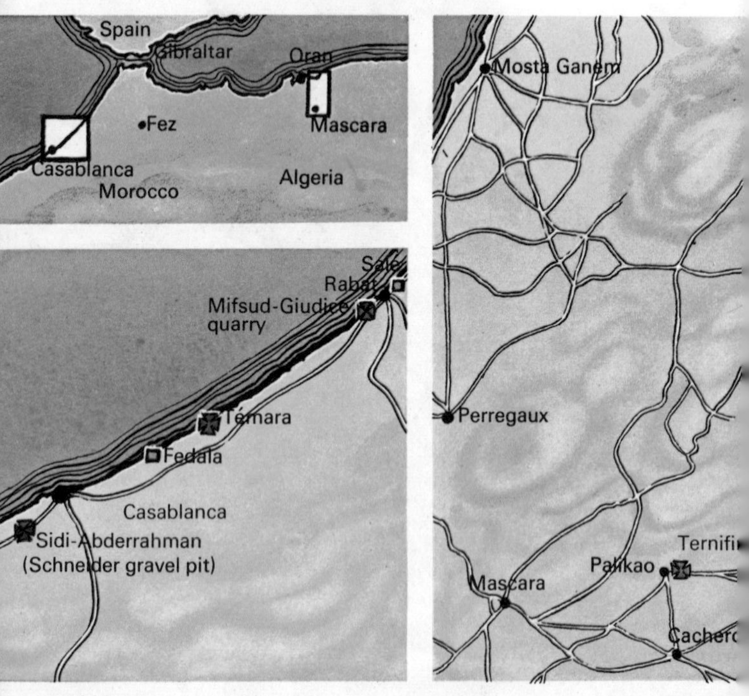

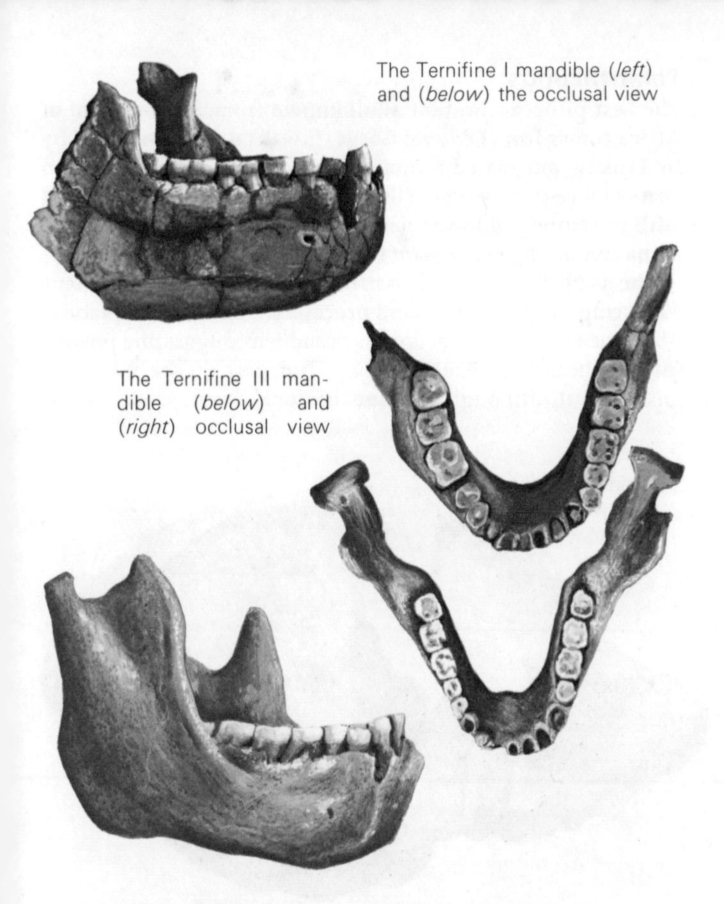

The Ternifine I mandible (*left*) and (*below*) the occlusal view

The Ternifine III mandible (*below*) and (*right*) occlusal view

The parietal bone is young, for the sutures were open at the time of death, and it shows the flattened contour that characterizes similar bones from Java and Peking. Its thickness is within the range of variation found in modern man.

Curiously enough the other North African pithecanthropine fossil remains have all been jaws, the best examples deriving from Casablanca and Rabat. The jaws are broken, and retain only a few teeth, but these tend to have the characteristics of those from both Ternifine and Peking. The presence of enamel collars on the molars, wrinkled biting surfaces and taurodontism, argue for their allocation to *Homo erectus*.

Telanthropus

The best pithecanthropine skull known from the continent of Africa comes from Olduvai Gorge. It was taken from Bed II by Dr Leakey, and dated radiometrically at about half a million years before the present. The skull was found to be associated with Developed Oldowan hand-axes and the fossil remains of a fauna containing many giant forms.

The skull is thick and has frontal flattening, a prominent brow ridge and small mastoid processes. The occipital plane is oblique and the occipital crest is prominent, suggesting powerful neck muscles. The face, base and part of the vault are missing but, although the find has only been provisionally

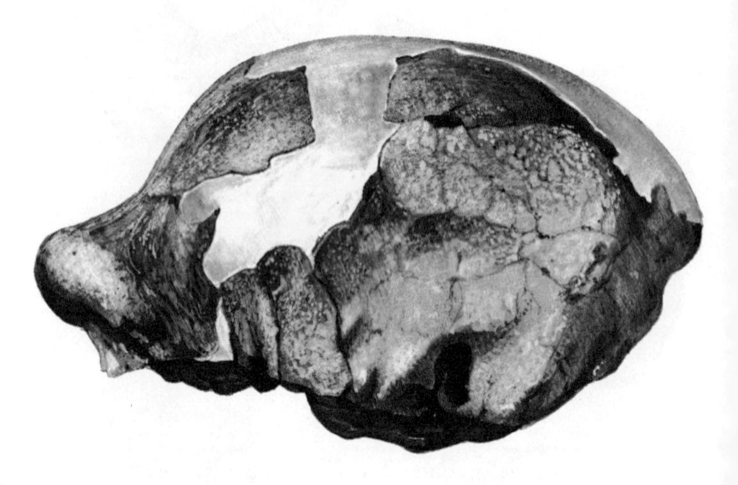

The Hominid 9 skull

described, the general likeness of this skull to those of Java and Peking is certainly most striking.

One other fossil jaw deserves mention since it has provoked much discussion. This mandible, together with a few other fragments, was found by Robinson at Swartkrans in the Transvaal. The body of the specimen was incomplete and contained five molar teeth, but unfortunately the whole jaw was somewhat crushed. There is thickening of the bone in the region of the third molar. Another part of a mandible has also been found at the same location and an isolated premolar tooth was found close to the better mandible.

At first the mandible was thought to belong to a new hominid genus which was named *Telanthropus*. Later this name was dropped and the material allocated to *Homo erectus*. This means that it is believed that there were pithecanthropines as well as australopithecines in the Transvaal during the early part of the Middle Pleistocene Period. The new diagnosis was made on the grounds of a reappraisal of the jaw and the teeth. Opponents of this view do not accept that this jaw shows any features in advance of those that are found in the australopithecines. Clearly much more material is needed before the dispute can be settled.

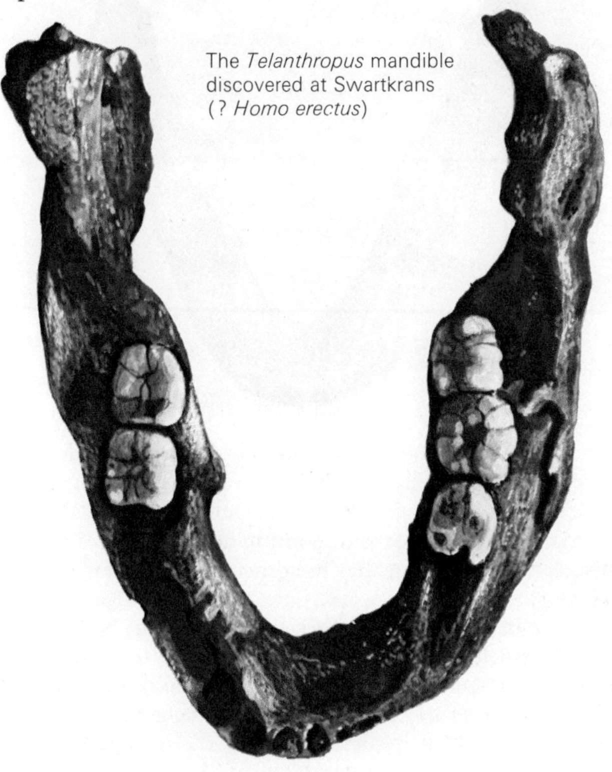

The *Telanthropus* mandible discovered at Swartkrans (? *Homo erectus*)

Heidelberg man

The Mauer sand-pit near Heidelberg produced one of the most perfect fossil jaws that has ever been found. This isolated jaw was found by a workman and shown to Otto Schoetensack in 1908. The exact layer in which the fossil was found is uncertain but it is attributed by most authorities to the Mauer sands of the First Interglacial Period. The mandible is large and robust, its rami being amongst the broadest known. The tooth row is rounded but the teeth seem to be small in

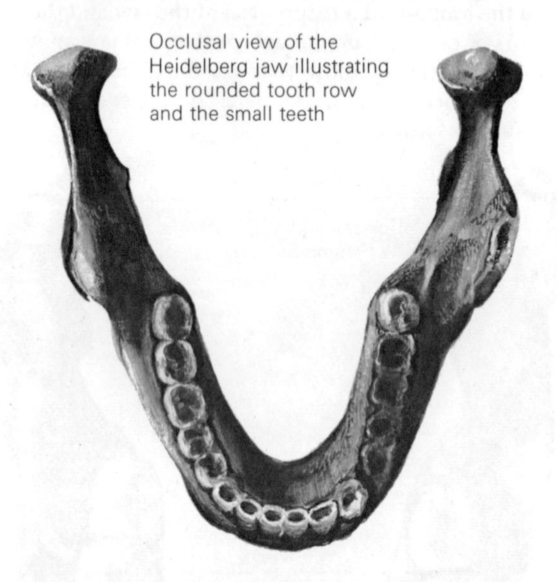

Occlusal view of the Heidelberg jaw illustrating the rounded tooth row and the small teeth

comparison with the size of the jaw. The teeth are rather modern in their structure since there is no sign of enamel collars (cingula) or secondary enamel wrinkling, although there is some degree of taurodontism of the molars. There are muscular markings but they are generally in keeping with the size of the bone.

This is an interesting jaw because it combines both ancient and modern features, that is, that of robust bone with progressive teeth. Because of these fundamental differences, it is not certain that this jaw belonged to *Homo erectus* and it has been suggested that it could represent the remains of a form intermediate between the pithecanthropines and the Neanderthalers, a group that flourished during the last glacial period.

The Middle Pleistocene hominids are perhaps the earliest widespread fossil group that fully merit the title 'fossil man'. They have been seen to occur from China to East Africa and from Java to the Mediterranean coast and in brain size they range from 850 cc.–1200 cc., while from the little that is known of their limb structure, it seems that they were probably modern walkers.

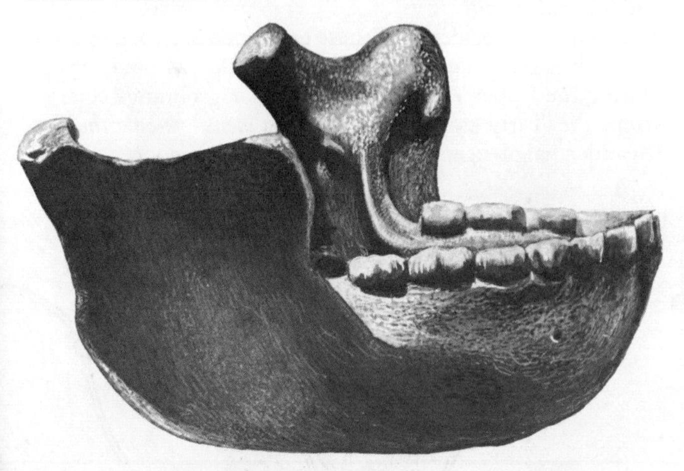

The Heidelberg mandible

In addition to the anatomical evidence, there is the cultural evidence of the use of hand-axes and cleavers as well as the control of fire. The length of the ash heaps suggests that Peking man knew how to kindle, as well as maintain, the flames on which they cooked the meat from the large mammals they hunted. The bones of large game recovered at this site also suggest that hunting was probably a co-operative venture necessitating some form of communication between members of the group. This would be a powerful incentive for the development of the beginnings of language, but there is no reliable means of telling from the brain size or the jaw structure whether the pithecanthropines or any other hominids were capable of articulate speech.

THE MODERN HUMAN PHASE

The early sapients

This period marked the beginning of the explosive expansion of mankind. Mastery of an economical method of locomotion, that is, walking or running leaving the hands free, plus the ability to manufacture weapons, tools and fire gave man the means to exploit his situation. Horizons were thus widened, game herds could be followed and the colder parts of the globe could be colonized.

Increasingly, evidence of these early men has been found in the rock shelters and caves of the Middle East and Europe. During the Upper Pleistocene two major groupings emerged from the early sapients, *Homo sapiens neanderthalensis* (Neanderthal man) and *Homo sapiens sapiens* (modern man).

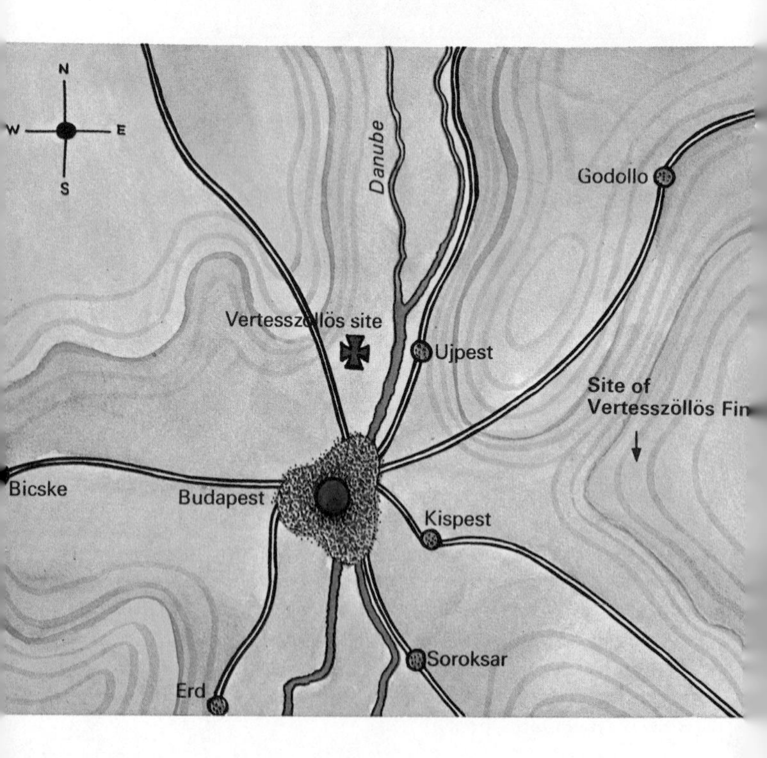

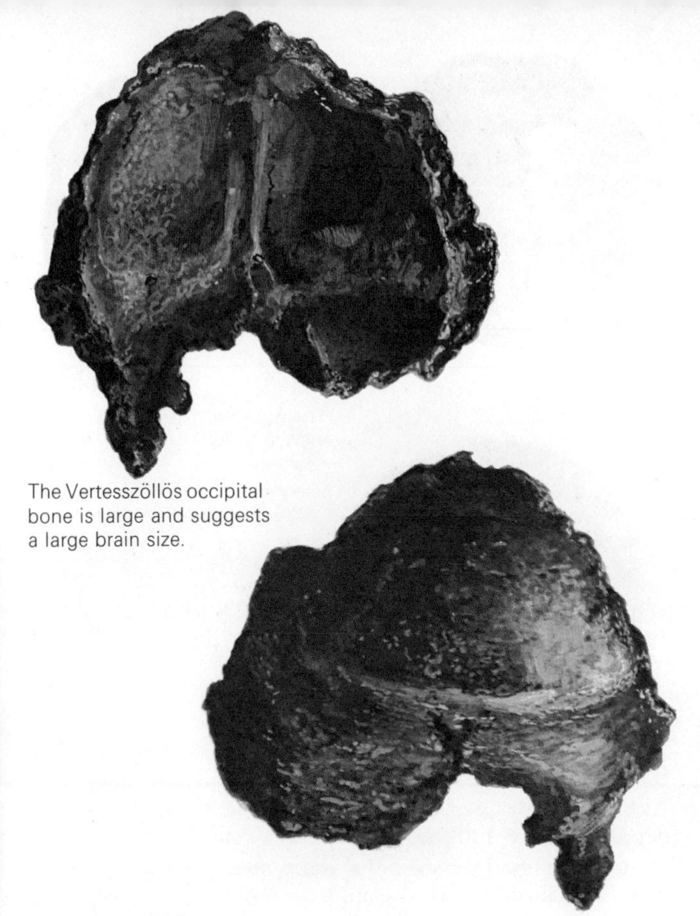

The Vertesszöllös occipital bone is large and suggests a large brain size.

Vertesszöllös man

At Vertesszöllös, near Budapest in Hungary, a site was found in 1965 that may prove to be one of the earliest known modern human living sites. An occipital bone was recovered from an occupation floor in deposits laid down during a warm phase in the Mindel glaciation. In addition, many small stone tools were found as well as a charred bone. This find may allow Vertesszöllös man to rival Peking man as the earliest known human to use fire. Preliminary assessment of its anatomy has led to the conclusion that it must have belonged to an early member of our own species, *Homo sapiens*.

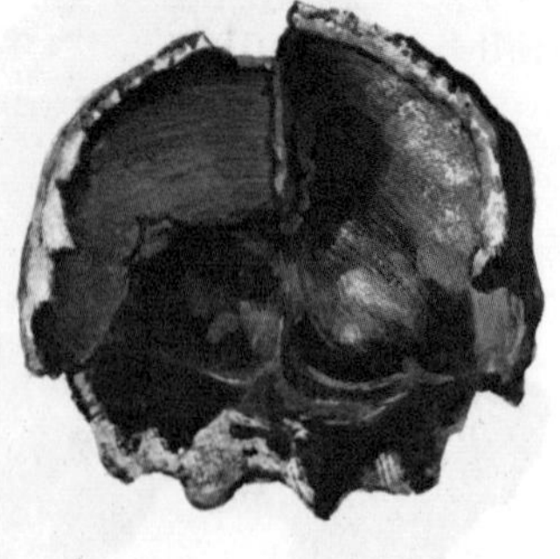

The Swanscombe skull, vertical view (*top left*), internal view (*above*) and occipital view (*left*)

Swanscombe

At Swanscombe, on the south bank of the River Thames, the deposits of the 100-foot terrace have long been known to contain the fossil remains of a warm inter-glacial fauna, and many well made hand-axes and flake tools. In 1935, 1936 and 1955, excavation in these beds produced three bones which form the back of a human skull; two parietals and an occipital bone. They fit together and belonged to one individual. The bones are thick and the skull is broad at the back, but from its general appearance and estimated capacity (1325 cc.) it is modern in shape. The key bone, the frontal, is missing and it is uncertain whether it had a prominent brow ridge or not. If we are to judge by another similar skull from Steinheim in West Germany, it is likely to be a lot more prominent than modern man, but less so than those of the so called 'classic' Neanderthalers of the last glaciation.

Steinheim skull

The Steinheim skull comes from a gravel pit near Stuttgart, in West Germany. It is about the same age as the Swanscombe skull from the Great Interglacial Period. The other mammalian fossils suggest a woodland environment since there were straight-tusked elephant, bear and wild oxen bones present in the deposit.

The skull lacks a jaw and has suffered considerable distortion but the vault and the face are complete on one side. It is a long narrow skull with a moderate brow ridge and a well rounded occiput, quite different from *Homo erectus* and yet not as exaggerated in its features as a Neanderthal skull. Unfortunately there are neither tools nor post-cranial bones known from the Steinheim deposits.

Interglacial riverine scene

The Middle East

The Middle East has provided further evidence of fossil man which suggests that intermediate forms existed between the classic Neanderthaler and the fully sapient forms discovered at more recent sites. Excavations at Mount Carmel in Israel produced a number of skeletons from two adjacent caves. The Skūhl cave contained ten skeletons while the Tabūn cave contained one skeleton and a jaw bone. The remains from the two caves differ in a number of important respects, and it is believed that, although the caves are adjacent, there is

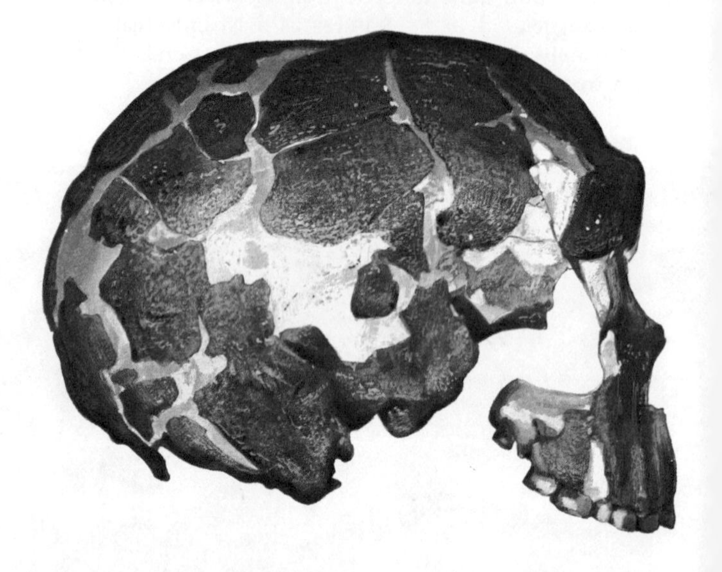

Lateral view of the Tabūn skull

probably a gap in time of about 10,000 years between the two fossil-bearing layers.

The Tabūn skeleton is that of an adult female who shows a number of Neanderthal features. The cranium is small and low-vaulted with a pronounced frontal ridge but there is no occipital chignon, or bun, formation. There is a prominent occipital torus, however, which denotes the presence of powerful neck muscles. The mandible is short with a stout body, broad rami and widely separated condyles. The mandibular symphysis is oblique and there is no chin. The teeth in the upper jaw are almost all present and indicate by their wear that coarse food was eaten. The molars have four-cusped square crowns but the third molars are triangular.

The limb bones of the Tabūn skeleton are short, thick and tend to be bowed, particularly the bones of the forearm. The spine is modern in its shape but the bones of the shoulder are coarse and strongly marked by muscle impressions. The pelvis is low and narrow with flattened pubic rami, a Neanderthal feature. Altogether the Tabūn skeleton is unquestionably Neanderthal in its affinities.

The Skūhl skeletons, although having some Neanderthal features, are more advanced. The best preserved skull is that of Skūhl V; the vault of this skull is high, the occipital region is full and rounded and the brow ridge well marked. The face is somewhat prognathic and the root of the nose is depressed in a fashion reminiscent of an Australian aboriginal skull. The mastoid process is large and the temporo-mandibular joint is modern in shape.

The mandibles are remarkable in that most of them have well marked chins, single mental foramina and modern condyles. It is of some interest that the jaws of Skūhl V show evidence of severe periodontal disease with abscess formation and the bony changes of pyorrhoea, although the teeth are not carious.

The limb bones of the Skūhl skeletons have mixed characters, the majority being like those of modern man. The feet, for example, are large, well arched and obviously capable of producing a good walking gait. The arm and leg bones are long and slender, in contrast to the short and curved bones of the Tabūn woman.

The mixture of features in the skeletons that were discovered led some authorities to speculate that these were the bones of a Neanderthal/sapient hybrid population. This theory is unlikely, however, because of the gap in time between the remains. It may be that the Tabūn peoples were the remains of a Neanderthal population who were replaced by a more advanced sapient people.

Zambia

From Africa, near Broken Hill in Zambia, a skeleton was recovered from a limestone hill that was being mined for lead and zinc ore. The hill had been tunnelled at its base and a cave was found that was filled with fossilized and mineralized bones. Together with the bones were some chert and quartz stone tools as well as a few bone tools. Fauna discovered at the site included large carnivores, zebra and several artiodactyls including an extinct buffalo.

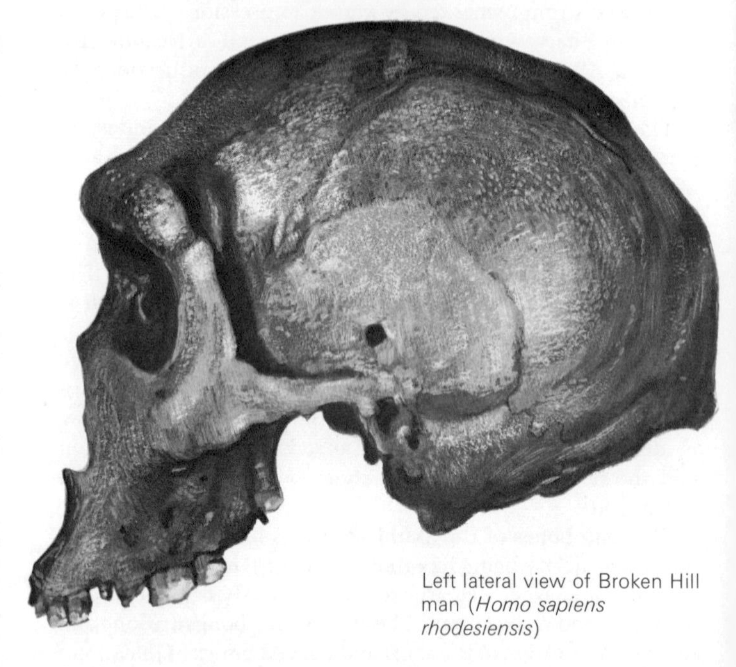

Left lateral view of Broken Hill man (*Homo sapiens rhodesiensis*)

It was believed initially that the bones were of differing ages because of variations in the mineral content of the specimens, but intensive radiometric investigations have indicated that the bones are in fact of the same age.

The bones are well preserved and once again show a combination of both Neanderthal and modern features. The skull is large with a massive brow ridge and flattened vault. The face is very large and the hard palate is broad. The teeth are worn and set in a U-shaped dental arcade and many show signs of decay and dental abscess formation.

The limb bones of Broken Hill man are long and stout but modern in shape. They indicate that this man was tall, powerfully built and upright in stance. A similar skull was found at Saldanha, near Cape Town, showing that this type of man was not an abnormal specimen, although classification of these remains is still controversial.

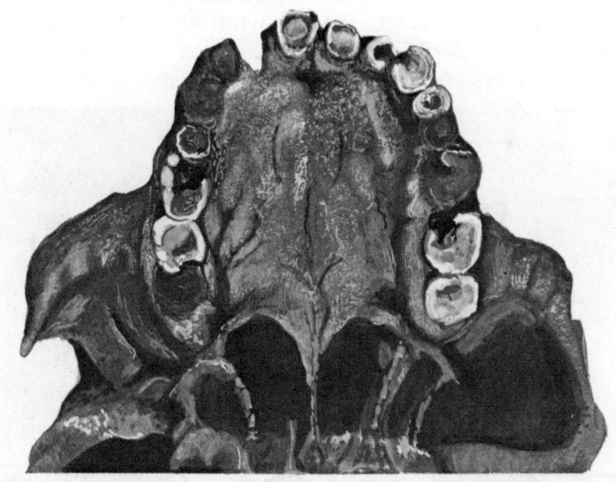

Basal view of the U-shaped dental arcade

In a recent assessment of the Swanscombe skull, it was suggested that Broken Hill man belongs to a group of intermediate types from which both the 'classic' Neanderthalers and the modern sapients arose. This 'spectrum of varieties' would include a number of other representatives from widely separated sites, for example, Europe (Steinheim), the Middle East (Skūhl & Tabūn), Yugoslavia (the Krapina remains), and Africa (Broken Hill and Saldanha). Thus it appears that there was an amalgam of forms present during the early part of the Upper Pleistocene, possibly derived from advanced *Homo erectus* stock, which diverged into the classic Neanderthalers of the Last Glaciation on the one hand, and the modern sapient forms on the other.

Neanderthalers

Neanderthal man is the colloquial name given to a group who bear resemblances to a famous skeleton found at Neander near Dusseldorf. Some years later several further skeletons of this kind were found in France, Italy, Belgium and the Middle East. In general they all date from the Würm Glaciation, the last glaciation to make the peninsula of Europe an icy and inhospitable place. Associated with Neanderthal remains, therefore, it is common to find the remains of a 'cold' fauna such as woolly rhinoceros, mammoth, reindeer, arctic marmot, wolf and bison. Neanderthalers lived in caves or rock shelters and made use of fire for warmth, light and probably cooking.

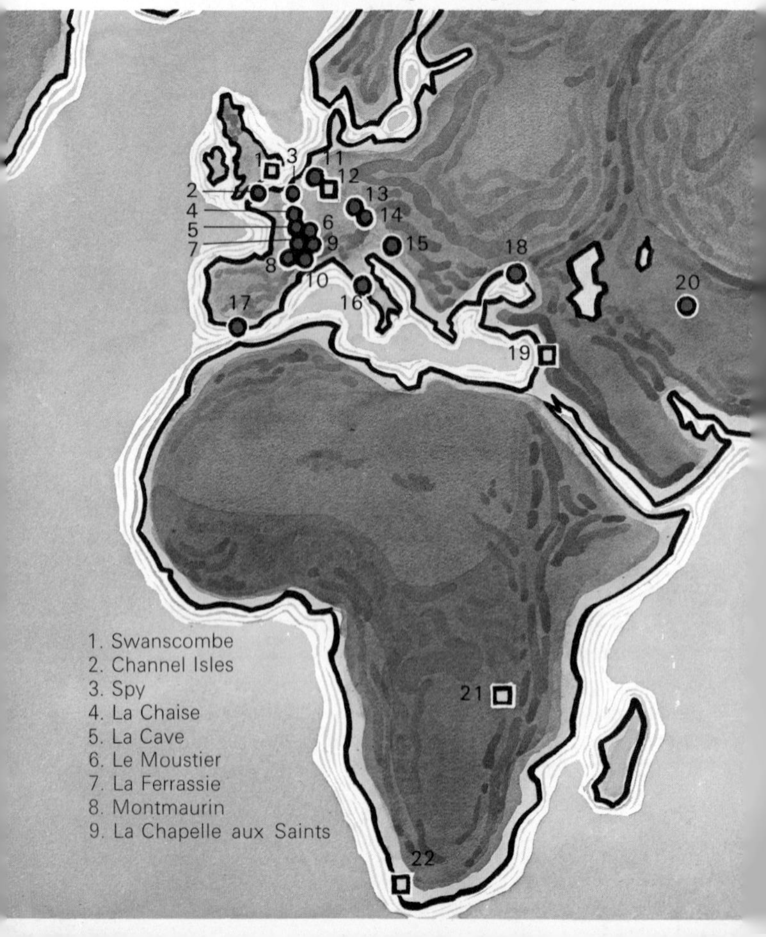

Perhaps the most striking feature of the Neanderthaler is the form of his skull. The general outline of the cranium is inflated by comparison with the *Homo erectus* or fully sapient skulls – in fact the cranial capacity of the classic Neanderthaler may be greater than that of modern man. None the less, the frontal region is flattened, the brow ridge prominent and the occipital region expanded by a curious chignon or bun-like swelling. The Neanderthal face is large and the nasal aperture broad, suggesting a large but flattened nose. The jaw is stout and chinless bearing teeth arranged in a horseshoe-shaped arcade. Neanderthal molar teeth frequently have enlarged pulp cavities, which is known as taurodontism.

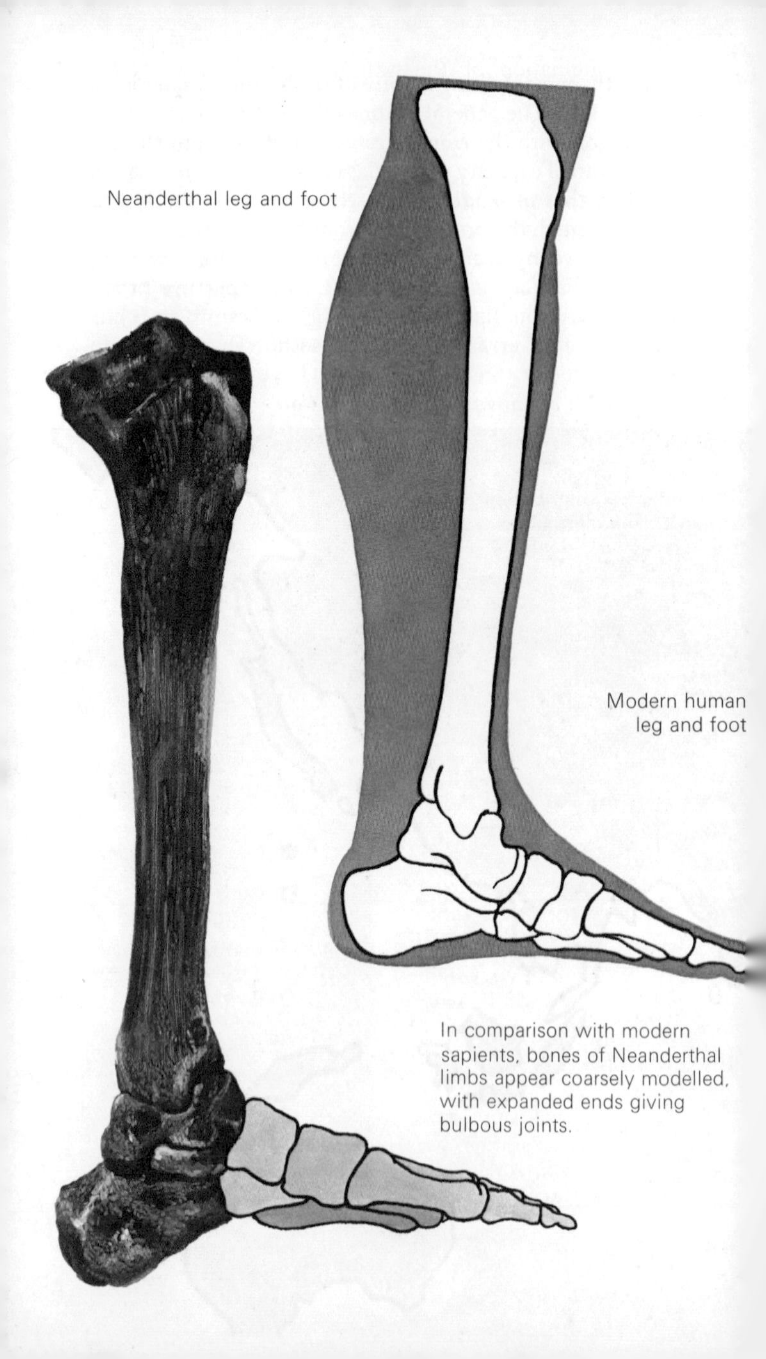

In comparison with modern sapients, bones of Neanderthal limbs appear coarsely modelled, with expanded ends giving bulbous joints.

From examination of the post-cranial bones of a number of classic Neanderthalers, it is clear that they were a short, powerful and thick-set race of men. This form of body contour is probably one of the reasons why they lasted as long as they did while the weather progressively deteriorated as the glaciation advanced. The short and broad body conserves heat better than the long thin body.

Neanderthal limb bones are similar to those of modern man but they have a number of distinctive characteristics. Muscle markings are often crudely inscribed and together with the curvature of many of the long bones they testify to the strength of these men.

The peculiarities of the Neanderthal post-cranial skeleton appear to have misled a number of earlier physical anthropologists into believing that Neanderthal man was habitually a stooping bent-kneed individual doomed to walk with a shuffling gait. This concept was based on a reconstruction of the skeleton from La Chapelle aux Saints, a man who has since been shown to be the victim of crippling arthritis.

There is no reason to believe that Neanderthal man could not stand, walk and run in a competently bipedal fashion, since his limb bones contain no evidence to the contrary. Original conceptions were based on the skeleton of a man crippled with arthritis and it is better to consider the bones of a healthy adult Neanderthaler such as La Ferrassie I.

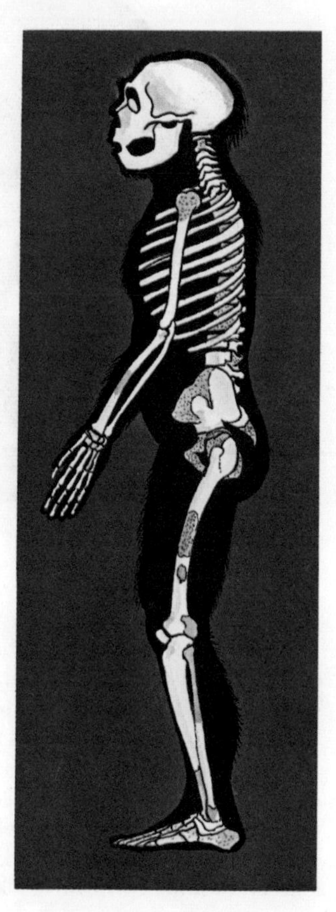

Neanderthal culture

The Neanderthalers of the last glaciation were culturally the most advanced group yet encountered; tool makers and fire users, cave dwellers and hunters. In addition, there is evidence from several Neanderthal sites of real cultural advance such as burial of the dead in places of safety from scavengers, the provision of 'grave goods' and the burial of the dead one's

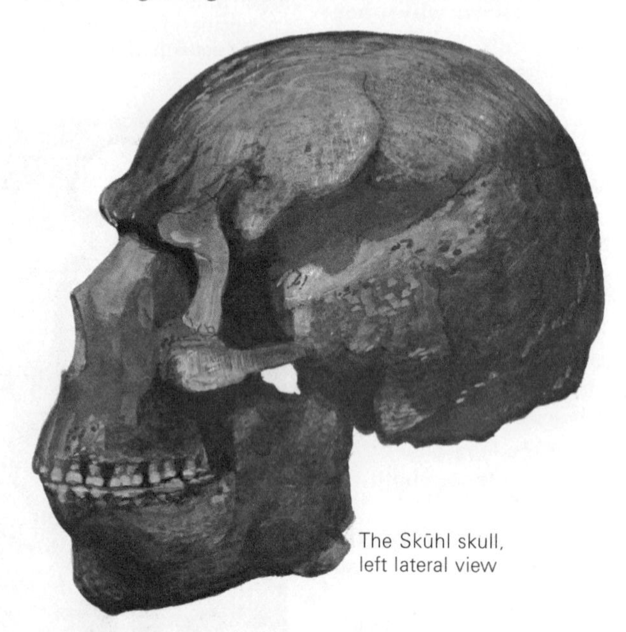

The Skūhl skull, left lateral view

personal belongings such as necklaces of perforated teeth.

It has been suggested that the beginnings of ritualism, mystic or religious practices can be shown from Neanderthal sites such as Teschik-Tasch and Monte Circeo. Remains have been found inside circles of objects placed with ritualistic precision.

When a cave in Italy, closed since Neanderthal times, was opened, cave bear remains were found with Neanderthal footprints in the hardened mud of the cave floor. Nearby was the stump of a torch and a sooty hand print on the wall, where this man had crouched in the darkness.

The Neanderthal race are the group of fossil men about whom more is known than any other. About sixty Neanderthal skeletons or parts of skeletons are known, ranging in distribution from Russia in the east to the Channel Isles in the west. In the south, North Africa and Iraq both have good Neanderthal sites. It is evident that as a people they were periglacial in their northern limit and were perhaps driven south

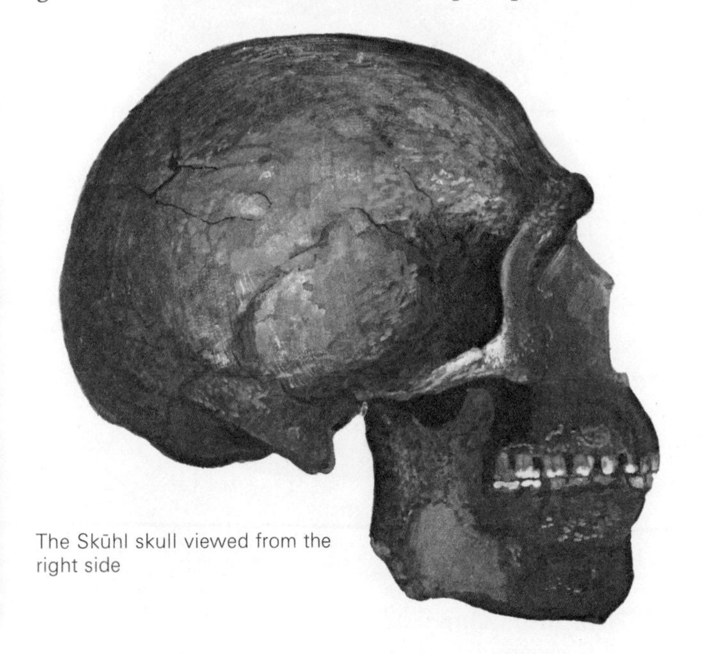

The Skūhl skull viewed from the right side

by the advancing ice. The Massif Centrale of France with its limestone structure and numerous caves probably housed a considerable population, particularly in the Dordogne.

Quite suddenly the Neanderthalers vanished from the scene. How and why they became extinct is still something of a mystery but they could have been wiped out by the invasion of more advanced men, or assimilated into the evolving population of modern man, or they may even have given rise to modern man. It is not impossible that all three processes took place to produce the same result.

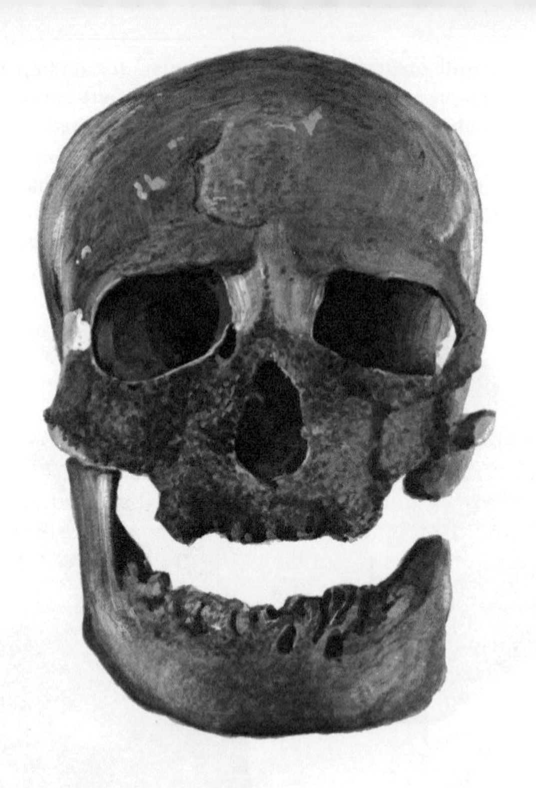

Cro-Magnon skull

Cro-Magnon Man

The final group of fossil men who display features that are noticeably different from those to be found in the wide variety of modern men alive today, are known as the Cro-Magnon race and they lived during the Upper Pleistocene Period. The type site of this group is in the Dordogne near Les Eyzies, in France. At this site, five adult skeletons were found as well as some infant bones.

The best skull of Cro-Magnon man is almost complete and is large, long and narrow. The face is broad and short and has a tall nasal opening and flattened rectangular orbits. The limb bones are long with good muscular markings which indicates tall stature and powerful limbs. Modern stance and gait are unquestionable. Many other sites in France, Germany, Italy, Great Britain and Czechoslovakia have provided similar skeletons.

The origin of these people is in some doubt because of the scarcity of the remains of their possible ancestors, for example Swanscombe or Steinheim. Similarly the relationship of these people to the Neanderthalers is unknown, however, at least some of the Cro-Magnon people were contemporaneous with the later Neanderthalers. It may be that Cro-Magnon man displaced, or over-ran and inter-bred with, the last of the Neanderthalers at the close of the last Glaciation and therefore the modern sapient stock emerged as the final product of hominid evolution. Subsequently geographical dispersion and genetic isolation has probably resulted in the variety of racial types in the world today. Several of these dispersed sapients have been discovered in sites, such as Mechta in North Africa, South Africa (Florisbad), China (Upper Cave, Peking) and Wadjak in Java.

Wadjak I skull

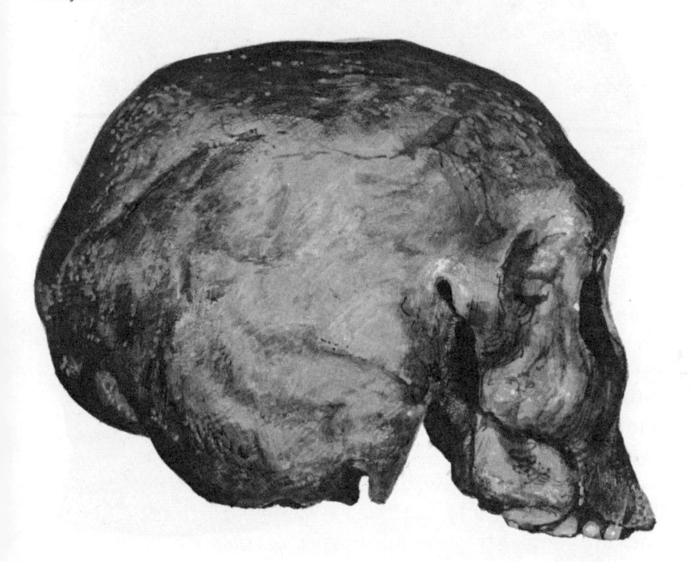

The successors

The men who in time succeeded the Cro-Magnon people are widespread and varied. Stone-age sites of a later period of time are numerous in eastern and southern Africa where preservation conditions are extremely good. In essence all the skulls and skeletons show no features which would lead any anatomist to suggest that they belonged to any group other than that of modern man.

Many attempts have been made to recognize in these fossil, and sub-fossil, remains from the Mesolithic Period, the forerunners of the modern human racial variants. 'Bushmanoid', 'Australoid', 'Boskopoid', 'Caucasoid' and 'Negroid' groups have been suggested, but some of these so-called 'types' are supported only by single specimens all of which are clearly sapient in form.

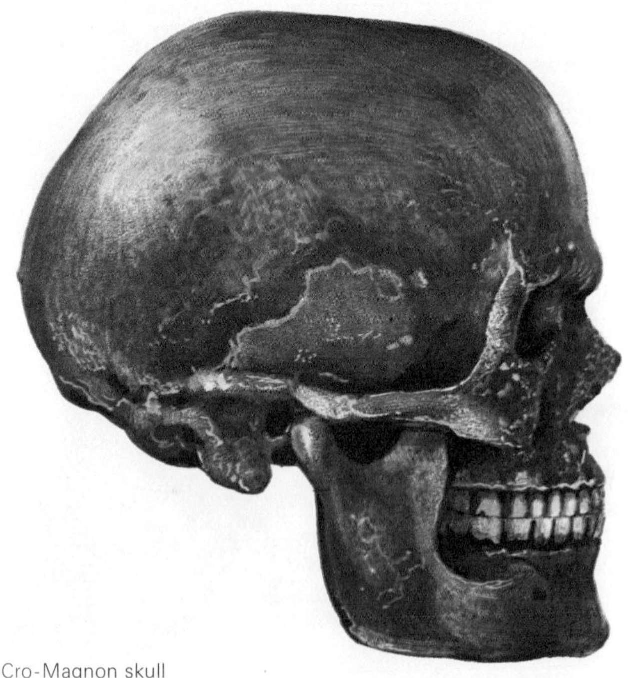

Cro-Magnon skull

Distinctions can quite easily be made between the peoples that follow the Cro-Magnon group. These do, however, become increasingly cultural and, thus, decreasingly anatomical.

The successors of Cro-Magnon man belong to the cultural phase that is known as the Neolithic Period. By this time, about 8000 years before the present, the climate, as well as the plants and animals, were much as they are today. The skeletons of Neolithic man are clearly sapient, but some features of the skull are of interest. Early human skulls tend to be rather long and narrow, and are called Dolichocephalic, while certain Neolithic skulls show the introduction of a more rounded shape, Brachycephalic. Neolithic skulls also tended to be rather thicker than those of present-day man.

The principal achievement of Neolithic man seems to be that he was able to break away from the constant necessity of moving to new hunting grounds, by beginning to produce food from edible plants and by taming useful animals. Prior to this breakthrough the numbers of people who could live in a given area or territory would have been severely limited by the concentration of game in the immediate neighbourhood. As this game was used up, family groups must have moved away to seek new hunting grounds and, therefore, the accent would have been on dispersal, rather than concentration of population. It would not have been possible to feed large numbers of people by hunting alone, since the hunting area would have had to be impossibly large to provide sufficient food. The beginnings of agriculture appeared in the Middle East by the growing of wheat and barley and the domestication of sheep, goats and cattle.

These first attempts at food production mark an important turning point in the history of mankind, since their success led to the possibility of co-operative social life in villages and, in turn, led to the division of labour so that those with particular skills could use them economically. In this way it is found that such implements as flint tools of Neolithic type, polished axes of greenstone and flint, were produced in 'axe factories' and distributed all over Britain. This manufacture and distribution would thus have led to the beginnings of trade. Clearly the Neolithic revolution can be said to have marked the real dawn of the processes of civilization.

Stone tools

One of the principal characteristics that distinguishes man from less evolved animals is his capacity to fashion a natural object and use it for a pre-determined purpose. Objects of stone, bone or horn which have been modified for a purpose are known as artifacts and can be identified by certain revealing features. Most artifacts are clearly tools or weapons which have been designed to give to the user capabilities that he does

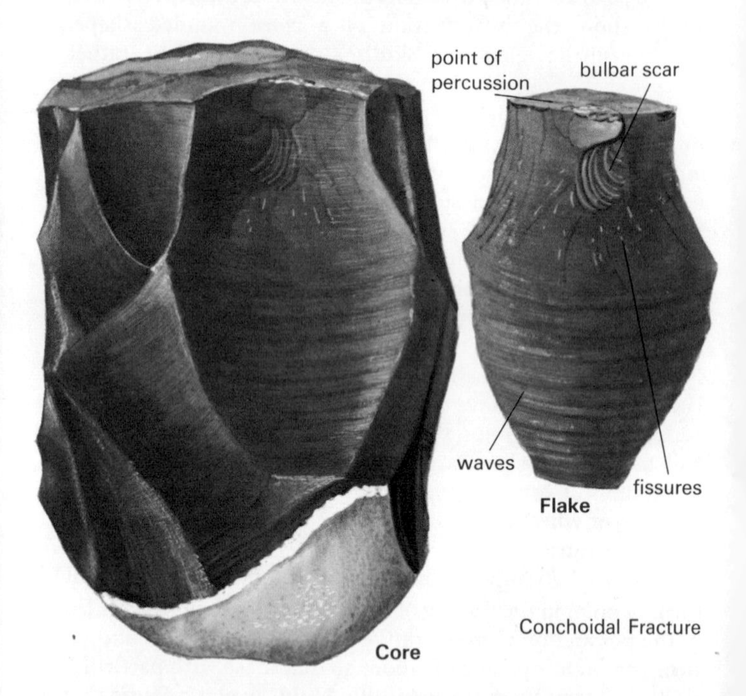

Conchoidal Fracture

not naturally possess. The evolutionary significance of these capabilities is enormous, allowing man, devoid of claws or fearsome teeth, to attack and kill his prey.

Stone tools betray their origins in several ways. Under natural conditions stones of flint may flake by reason of changes of temperature, frost action or extremes of heat. The flakes thus formed are rounded with concentric ripples and may cover a nodule in random fashion. When a flint nodule is deliberately struck, the point of impact forms the apex of a cone when the nodule shatters. Should the blow be directed at the edge of the nodule a half cone or bulb of percussion will be left and the negative bulb will be shown on the flake. The shape of the flake, or its parent core, may be altered to form a primitive blade, scraper or axe-head. The edge may be improved by delicate retouching. Evidence of transportation may be present and worked flints may occur in a deposit where geologically flint is foreign.

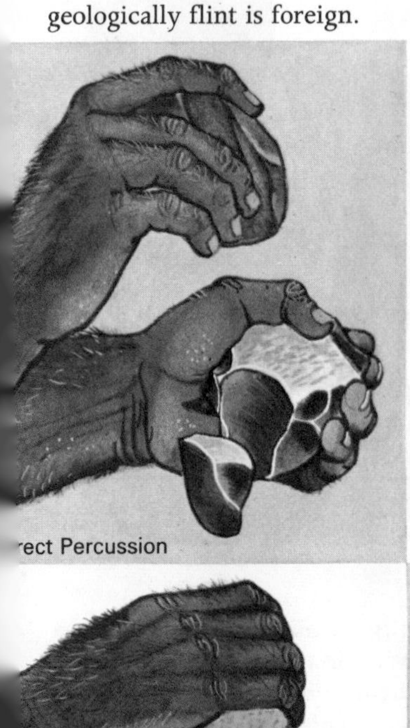

ect Percussion

Pressure Flaking

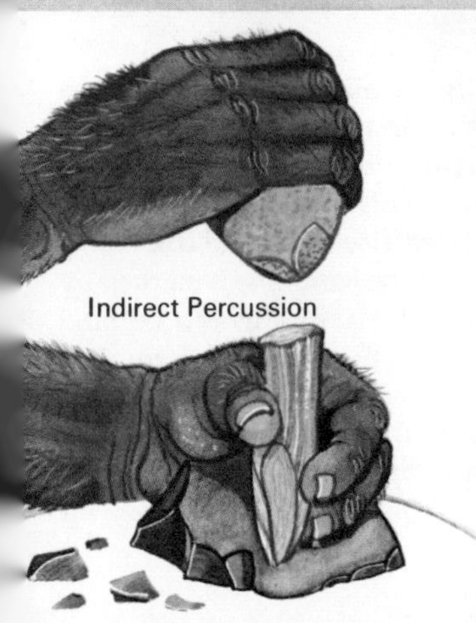

Indirect Percussion

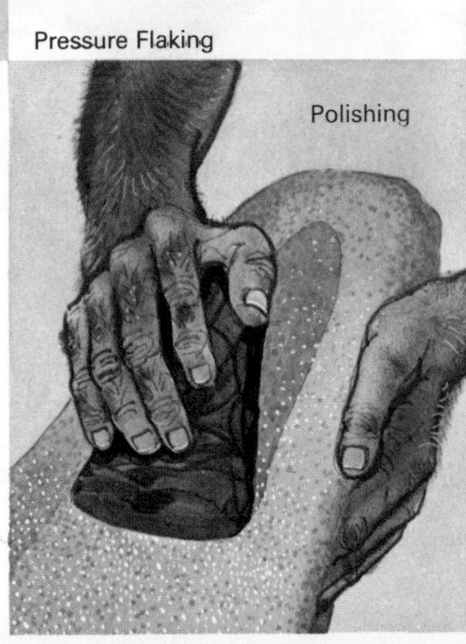

Polishing

Early Palaeolithic tools

The tools of this Period are simple, consisting of stones that have been flaked in one or two directions only. Many tools of this type have been recovered from Olduvai Gorge, and they are said to belong to the *Oldowan* culture. Their undoubted association with the hominid remains recovered from this site makes these hominids the earliest known tool-makers.

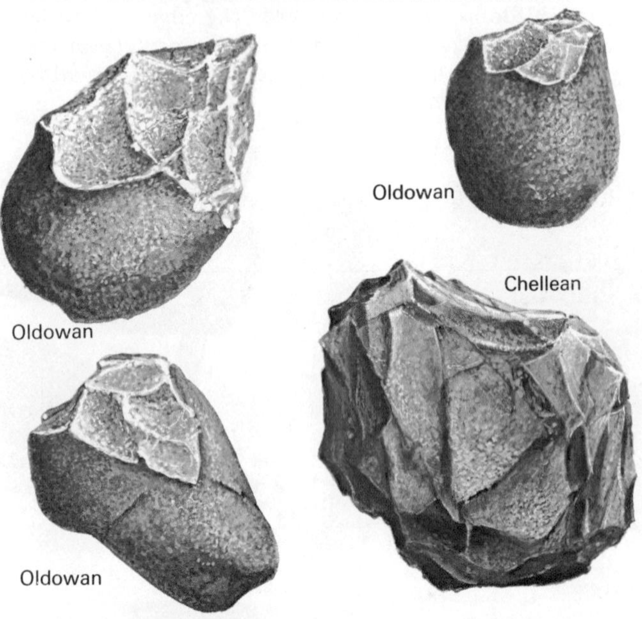

If the flaking process is continued all round the edge of a stone, a two-faced implement with an irregular border can be produced. Crude hand-axes of oval or pear-shaped outline, are termed *Abbevillian* axes. Tools of this type showing rather better workmanship may be termed *Chellean* hand-axes.

The third stage in the evolution of the hand-axe is marked by the appearance of the *Acheulean* hand-axe. This is a pointed tool and has straight edges retouched by delicate flaking. The surface has been trimmed by the removal of small, thin flakes and this contrasts strongly with the crude flaking of the previous Abbevillian and Chellean tools.

It is thought that this technological advance was brought about by the introduction of wood or bone as a flaking implement since the use of a hammer stone for flaking produces large crude flakes. Acheulean tools are widespread in their distribution and tend to vary locally in the details of their manufacture. Acheulean oval hand-axes with an S-shaped border tend to be common in western Europe.

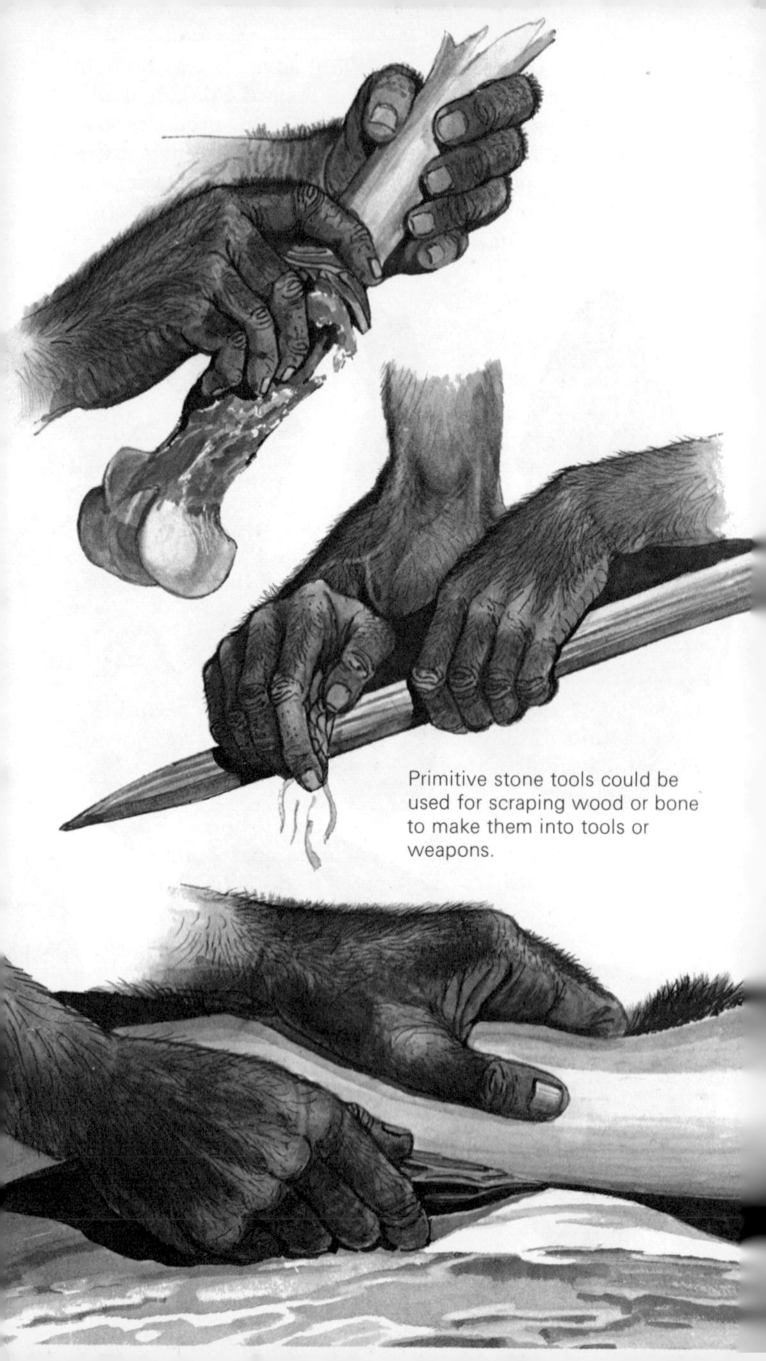

Primitive stone tools could be used for scraping wood or bone to make them into tools or weapons.

The manner in which the hand-axe was held

The Middle Pleistocene cultures

In the southern hemisphere, the evolution of stone tool culture proceeded rather differently. Some of the earliest known implements from Asia are the Choukoutienien chopper tools from Peking, which were used by *Homo erectus pekinensis* in the Middle Pleistocene Period. The tools tend to be coarse flakes shaped to form cleavers made of imported quartz and greenstone. Similar industries are known from India (the *Soan* culture) and from Java (the *Patjitanian* culture).

Utilized flake industries are also known from Europe. The *Clactonian* industry consists of tools struck from flint cores by making use of an anvil stone. The flakes which result are characterized by a prominent bulb of percussion and a wide angle between the bulb face and the striking platform.

In eastern and central Europe, a similar flake culture known as the *Tayacian* is found. Some authorities consider that this early palaeolithic flake culture developed into the *Mousterian* culture, typical of Neanderthal man, and the prepared core *Levalloisian* tradition.

Levalloisian tortoise-core (*left*) and flake (*above*)

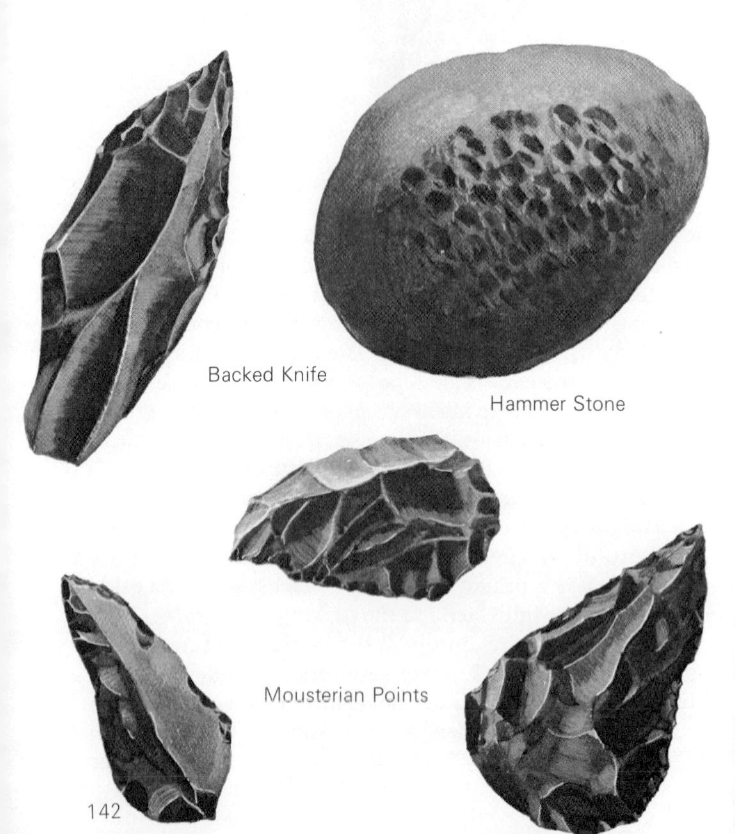

Backed Knife

Hammer Stone

Mousterian Points

Levalloisian tools

The *Levalloisian* flake is a carefully and skilfully made tool. A nodule of flint must be prepared to form a core from which the flake can be struck. The prepared core resembles a tortoise that has got on to its back. The flake is struck from the flatter side. Such a flake will have one surface showing much evidence of flaking whereas the other surface will show a bulb of percussion and a plain conchoidal fracture surface. The beauty of the method is that it is possible to produce thin flakes with very sharp edges that need little or no retouching. The butt of the flake tends to be faceted since it cuts across the prepared part of the core.

Mousterian culture

The *Mousterian* culture represents the tool-making tradition of the Neanderthal peoples since it is commonly associated with Neanderthal remains. The implements are often flakes that have been retouched to form side-scrapers, hammer-stones and some hand-axes. The Neanderthalers were proficient hunters, killing cave bear, mammoth and woolly rhinoceros. It seems likely that the carcass of a kill was cut up on the spot using these hand tools because a preponderance of animal bones found at Neanderthal sites come from limbs.

A Mousterian hide scraper

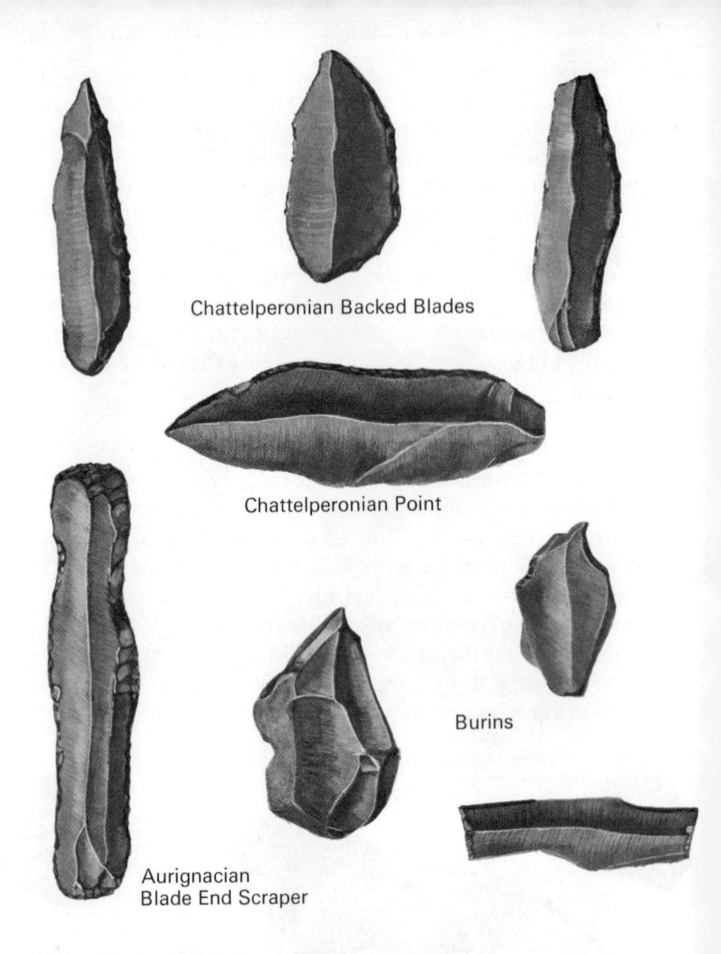

Late Palaeolithic tools

The Late Palaeolithic tools of the latter part of the Last Glaciation tend to follow the Mousterian culture at many sites in Europe and suggest that they were brought by the people who succeeded the Neanderthalers. The new culture consists of finely made blades and specialized engraving instruments known as burins. These tools usually have a narrow chisel edge and were used for working soft stone, wood, antler or bone. Many types of burin are recognized.

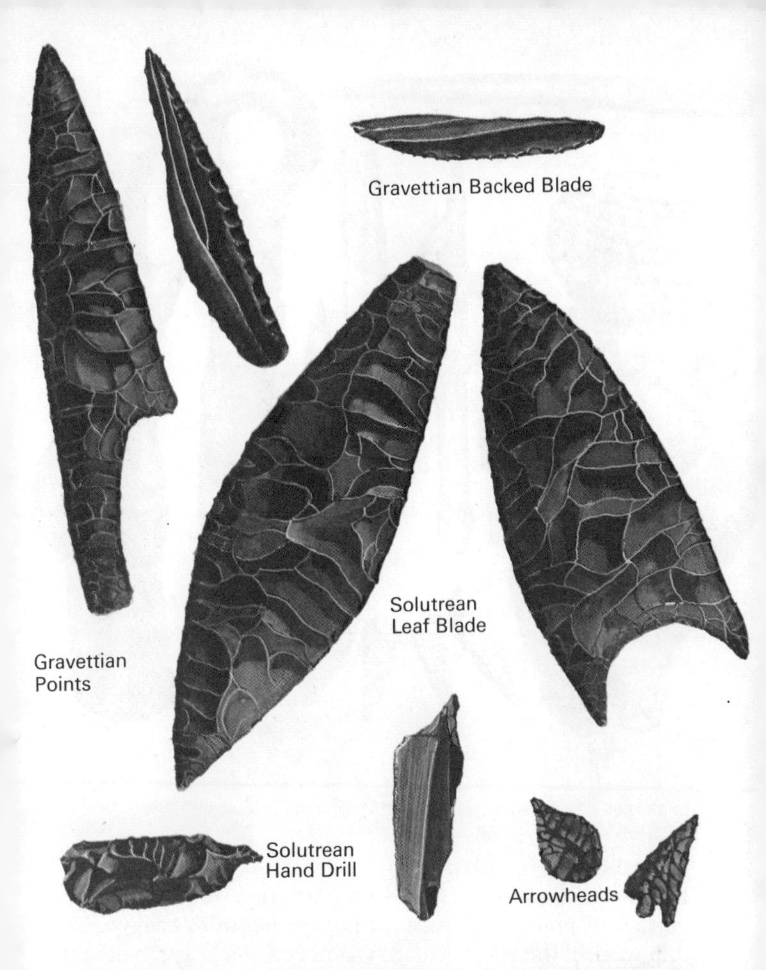

The earliest form of Late Palaeolithic culture is the *Chattelperronian,* an industry characterized by a flint blade with a blunt back. Later, the *Aurignacian* culture marked the beginning of the working of bone pins and spear points.

At this point in the history of culture, expansion began rapidly. New techniques and new ideas spread widely with migrating groups following the game herds. *Gravettian* backed blades and beautifully made *Solutrean* laurel-leaf blades begin to appear together with early arrowheads.

Examples of the Aurignacian and Magdalenian culture

Magdalenian culture

During the final part of the Last Glaciation, the *Magdalenian* culture of France and western Europe began to emerge. The flint work of the Magdalenians was not particularly remarkable but they were masters of the working of bone and antler. Among the many Magdalenian bone tools or weapons that have been discovered are included harpoons, spear-points with barbs, 'link-shaft' spears, eyed needles and spear-throwers.

The Magdalenian culture of the Late Palaeolithic has much in common with that of the Eskimo people for they have both had to cope with similar problems which they have managed to solve in similar ways.

Cave paintings

Only useful tools have been so far considered but perhaps the most remarkable records of fossil men are the cave paintings and sculptures that have been found in a variety of sites, more particularly in southern France and Spain. The earliest works of art are attributed to the Aurignacian Period and in this, and the subsequent Solutrean and Magdalenian Periods, men learned to draw, paint, engrave, sculpt in bas-relief and sculpt in the round. Much of the work that has survived is of outstanding quality and beauty, such as carvings of animal heads, small Venus-like figures, and the famous cave paintings of Lascaux, Altamira and many other sites. It is difficult to understand why these works of art were produced, often in the depths of dark caves. Many of the paintings are of animals, probably those hunted, and perhaps the most likely explanation is that these men believed in a form of 'hunting magic', although no single explanation can account for everything.

Magdalenian cave paintings

Metal tools

The first metals that were used to make tools were those that sometimes occur as metal in the natural state. Later it was discovered that metal could be produced by the smelting of ores and this led to an extension of its use.

Copper is a soft metal that can be shaped by hammering, or cast by heating and pouring into prepared moulds. As a material for tool-making it has drawbacks because of its softness, but, by adding about ten per cent of tin to copper, bronze can be made which is a hard alloy. Pure iron is fairly soft but cannot be beaten cold. If it is beaten when red-hot, it can be more easily shaped. This process is known as forging.

The first metals used by man were copper, tin, iron, lead, silver and gold.

Primitive metal blades are not sharper than new flint tools, but they are less brittle than flint blades and when the edge becomes dulled metal edges can be sharpened by grinding on a flat stone. With soft metal tools this must have been a frequent and tedious business. It was quite possible that for this reason stone tools were not finally superseded until it was discovered that iron, heated slowly in a charcoal fire, was considerably hardened and in fact an early type of steel was formed.

Cooling hot metal rapidly in water (quenching) hardens it further, while reheating and cooling slowly (annealing) softens it. By judging these processes carefully the properties of a metal can be controlled and tempered blades produced.

Copper and bronze

The first metal that was used by early man was copper, but very little is known about the Copper Age or the men who pioneered this major technical advance. Later, the beginning of the Bronze Age was marked by the influx of rounder headed peoples into Europe and the British Isles. These men were responsible for the round burial mounds that can be found at many sites in Britain, which succeeded the longer mounds made by earlier more dolichocephalic people. This results in the old archaeological saying, 'Long heads, long barrows; round heads, round barrows'.

The workers of bronze were skilful and artistic and many beautiful examples of their work have been discovered, such as swords, shields, pins, buckles and many other objects.

Copper can be cast by heating and pouring into prepared moulds.

Hill forts

Later Iron Age people were responsible for building a considerable number of hill-top forts or fortified encampments in many parts of Britain. These forts usually consisted of concentric series of earthworks that were probably surmounted by palisades. These were situated on prominent sites that commanded the surrounding countryside. Many of these forts have survived to the present day including a fine example known as Maiden Castle in Dorset.

Thus, from the very beginning of his evolution, *Homo sapiens* has shown a high degree of skill. Between the foundation of ancient civilization and the present one, however, lies a long period of continuous, but nevertheless accelerating, technical change.

Iron Age fort, Maiden Castle in Dorset

Summary of evolutionary changes

Every family has its family tree and the more famous the family perhaps the further back the tree can be traced in recorded history. This approach to tracing one's ancestors, that of the genealogist, has been used in the past by anthropologists and has led to many fruitless controversies about how to place fossil finds in 'ancestral' relationship to each other. It should always be remembered that every fossil find was a member of a large breeding population, a population subject to the laws of anatomical variation. Thus to say that fossil *A* was 'ancestral' to fossil *B* can never be true in individual terms. Evolutionary changes must always be considered in terms of successive populations showing anatomical trends that result from the interaction of their genetic structure with the process of natural selection.

The Hominization Process

Structural Groups	Taxonomic Groups	Anatomical Features
Modern sapients	*Homo sapiens sapiens*	Modern teeth and jaws; chins present. Large brains (Range 1,000–2,000 cc., mean 1,300 cc.). Striding bipedalism. Power and precision grips. (Stone, bone and later metal tools, use of fire)
	Homo sapiens neanderthalensis	Less modern teeth; no chins present. Large brains (Range 1,200–1,600 cc.). Good bipedalism. Power grip, perhaps precision grip. (Stone tools only, use of fire)
Late human phase	*Homo erectus*	Less advanced, larger teeth; no chins. Smaller brains (Range 750–1,200 cc.). Probably good bipedalism. Grip not known. (Crude stone tools, use of fire)
Early human phase	*Homo habilis*	Less advanced, larger teeth; no chins. Small brains (Approximately 670 cc.). Advanced bipedalism. Power grip. (Early stone tool makers)
Pre-human phase	*Australopithecus africanus*	Hominid dentition. Very small brains (500 cc.). Early bipedalism. Grip not known. (Tool users, possibly early tool makers)
Early hominids	*Ramapithecus punjabicus* *Kenyapithecus wickerii* *Kenyapithecus africanus (?)*	Hominid dental features, nothing else known

The major anatomical trends discernible through hominid evolution consist of changes in the feet, legs, pelvis and lower vertebral column that culminated in striding bipedalism, as well as the development of manual skills through changes in the hands, arms and shoulders, and reduction of the size of the teeth, jaws and face accompanied by expansion of the brain. All of these trends are linked to each other, but, individually, they proceeded at variable rates. The result is termed 'mosaic evolution', which means parts of the body evolved at differential rates. Changes in the teeth and the locomotor system began early, whereas expansion of the brain lagged behind at first then increased rapidly.

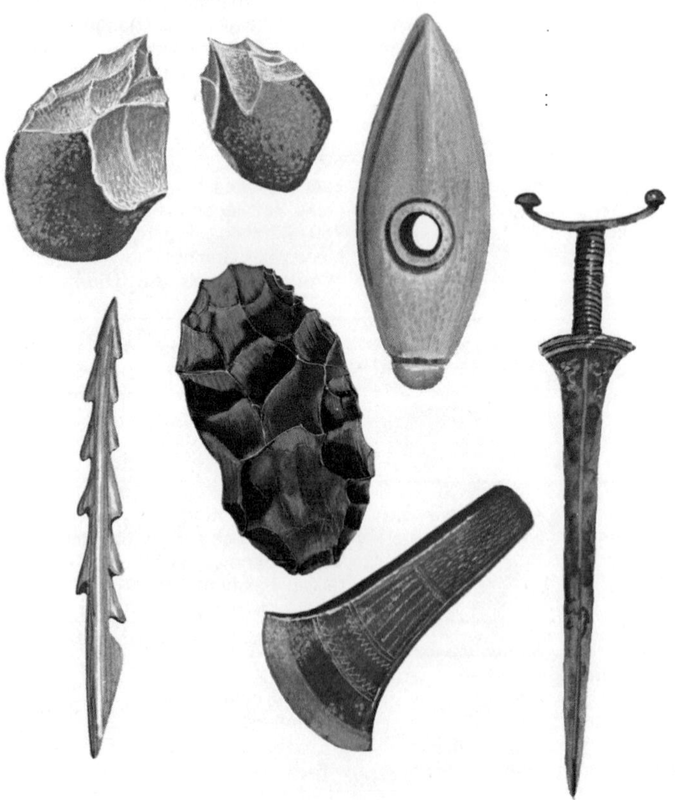

If you have been searching this book for a neat and simple explanation of the detailed evolutionary changes that have resulted in the production of modern man, you are probably disappointed. In truth there is no simple explanation available. The fossil record only allows a glimpse of the structure of early man at a few rather uncertain times during his evolutionary history. Anatomists make what they can of the few precious finds they have by close examination and analysis; every new find brings more knowledge and very often more problems, but this is the way of science.

The best synthesis that can be offered, in honesty, is to indicate the major structural phases leading towards hominization, with the warning that individual known fossils within those groups are by no means necessarily ancestral to each other.

What is known of the evolution of man is still but a fraction of what there is yet to learn.

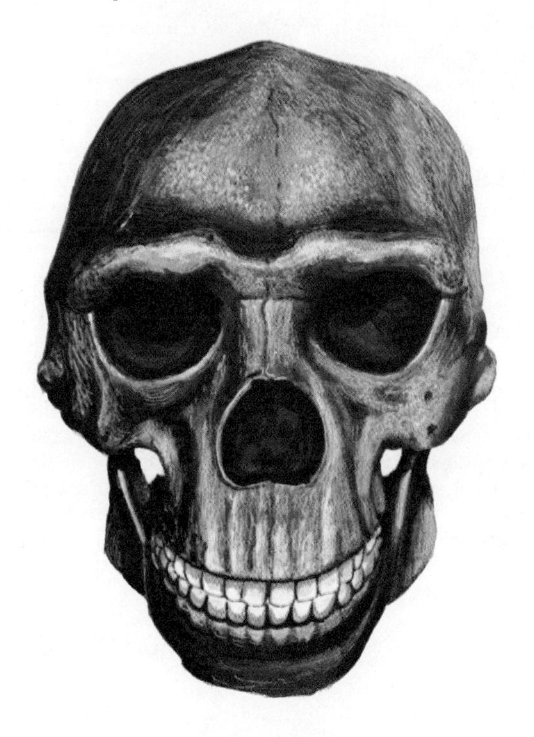

BOOKS TO READ

Dating the Past by F. E. Zeuner. Methuen, London, 1958.
Early Man in East Africa by Sonia Cole. Macmillan, London, 1958.
Environment and Archeology by Karl Butzer. Methuen, London, 1965.
Fossil Evidence for Human Evolution by W. E. Le Gros Clark. Chicago University Press, 1964.
Framework for Dating Fossil Man by K. P. Oakley. Weidenfeld and Nicolson, London, 1964.
Glacial and Pleistocene Geology by R. F. Flint. Wiley, Chichester.
Guide to Fossil Man by M. H. Day. Cassell, London, 1965.
History of the Primates by W. E. Le Gros Clark. British Museum (Natural History).
Inorganic Raw Materials of Antiquity by A. Rosenfeld. Weidenfeld and Nicolson, London, 1965.
Man and the Vertebrates by A. S. Romer. Penguin (Pelican), two volumes.
Man the Tool Maker by K. P. Oakley. British Museum (Natural History), 1963.
Origins of Man: Physical Anthropology by J. Buettner-Janusch. Wiley, Chichester, 1966.
Races of Man by Sonia Cole. British Museum (Natural History), 1963.
The Pleistocene Period by F. E. Zeuner. Hutchinson, London, 1959.

PLACES TO VISIT

The British Museum, (Natural History), London S.W.7. (Extensive Evolutionary exhibits)
The British Museum, London W.1. (Stone tools but no fossils)
Musée de L'homme, Palais de Chaillot, Paris -16e. (Large collection of human material and stone tools)
Musée des Eyzies, Les Eyzies, Dordogne, France. (Original bones, stone tools and cave art; nearby many prehistoric sites open to the public)
The Transvaal Museum, Pretoria, Republic of South Africa. (Many australopithecine fossils kept here)
The National Museum of Kenya, Nairobi, Kenya. (An extensive collection of fossils and tools from East Africa)
The University Museum, University of Pennsylvania, Philadelphia, 4, U.S.A.
Peabody Museum, Harvard University, Cambridge, Mass., U.S.A.
The American Museum of Natural History, New York, U.S.A.
The United States National Museum, Washington D.C., U.S.A.

INDEX

Figures in bold type refer to illustrations.

SOME OTHER TITLES IN THIS SERIES

- Arts
- Domestic Animals and Pets
- Domestic Science
- Gardening
- General Information
- History and Mythology
- Natural History
- Popular Science

Arts
Antique Furniture/Architecture/Clocks and Watches/Glass for Collectors/Jewellery/Musical Instruments/Porcelain/Victoriana

Domestic Animals and Pets
Budgerigars/Cats/Dog Care/Dogs/Horses and Ponies/Pet Birds/Pets for Children/Tropical Freshwater Aquaria/Tropical Marine Aquaria

Domestic Science
Flower Arranging

Gardening
Chrysanthemums/Garden Flowers/Garden Shrubs/House Plants/Plants for Small Gardens/Roses

General Information
Aircraft/Arms and Armour/Coins and Medals/Flags/Guns/Military Uniforms/National Costumes of the world/Rockets and Missiles/Sailing/Sailing Ships and Sailing Craft/Sea Fishing/Trains/Veteran and Vintage Cars/Warships

History and Mythology
Age of Shakespeare/Archaeology/Discovery of: Africa/The American West/Australia/Japan/North America/South America/Myths and Legends of: Africa/Ancient Egypt/Ancient Greece/Ancient Rome/India/The South Seas/Witchcraft and Black Magic

Natural History
The Animal Kingdom/Animals of Australia and New Zealand/Animals of Southern Asia/Bird Behaviour/Birds of Prey/Butterflies/Evolution of Life/Fishes of the world/Fossil Man/A Guide to the Seashore/ Life in the Sea/Mammals of the world/Monkeys and Apes/Natural History Collecting/The Plant Kingdom/Prehistoric Animals/Seabirds/Seashells/Snakes of the world/Trees of the World/Tropical Birds/Wild Cats

Popular Science
Astronomy/Atomic Energy/Chemistry/Computers at Work/The Earth/Electricity/Electronics/Exploring the Planets/The Human Body/Mathematics/Microscopes and Microscopic Life/Undersea Exploration/The Weather Guide